The Renaissance of American Steel

The Renaissance of American Steel

LESSONS FOR MANAGERS IN COMPETITIVE INDUSTRIES

ROGER S. AHLBRANDT
RICHARD J. FRUEHAN
FRANK GIARRATANI

OXFORD UNIVERSITY PRESS
New York Oxford
1996

Oxford University Press

Oxford New York
Athens Auckland Bangkok Bombay
Calcutta Cape Town Dar es Salaam Delhi
Florence Hong Kong Istanbul Karachi
Kuala Lumpur Madras Madrid Melbourne
Mexico City Nairobi Paris Singapore
Taipei Tokyo Toronto

and associated companies
Berlin Ibadan

Published by Oxford University Press, Inc.
198 Madison Avenue, New York, New York 10016

Ahlbrandt, Roger S.
The renaissance of American Steel : lessons for managers in competitive industries /
Roger S. Ahlbrandt, Richard J. Fruehan, Frank Giarratani
p. cm.
Includes bibliographical references and index.
ISBN 0–19–510828–0
1. Steel industry and trade—United States. 2. Corporate turnarounds—United States.
I. Fruehan, R.J. II. Giarratani, Frank. III. Title.
HD9515.A594 1996
338.4'7669'0973—dc20 96–25415

1 3 5 7 9 8 6 4 2

Printed in the United States of America
on acid-free paper

This book is dedicated

to the people dearest to us:

Julie, Bonnie, and Kathy

Preface

This book is about succeeding in an industry that is experiencing strong competition and technological change. Not so many years ago, fingers were pointed at the American steel industry as a prime example of failure in the face of international competition. So dramatic was the industry's decline that it was sometimes seen as symptomatic of larger failure in our economic system. Today, the American steel industry truly has been transformed. American firms produce quality products in efficient plants, and they are profitable.

We set out several years ago to understand this transformation and learn from it. Our intuition was simple enough. If you want to find out what it takes to be competitive in a global economy—if you want to know what management practices actually work—turn to an industry where restructuring has succeeded. Learn about the firms that helped turn that industry around, and seek lessons from their success that generalize broadly. We believe that readers of this book will find, just as we did, that the generalization carries lessons from steel to industries as far afield as financial services, wholesale and retail trade, utilities, computers, and semiconductors.

In the steel industry, two types of firms have been responsible for the transformation: large integrated producers that were tagged as failures in the 1980s but came back strong, and their small insurgent competitors, so-called minimills. We went to both types of firms to learn from their experience. We visited plants, talked to workers and managers, and interviewed the leaders who initiated change and helped make that change successful.

Our analysis did not stop there. We went to steelmakers abroad too. Restructuring in this industry has been worldwide, so it stands to reason that the basis for competitive success should cut across national boundaries. If the insights that we gained on the basis of experience in the United States could be generalized, we reasoned that restructuring here would find a counterpart in Europe and Japan.

The examples we draw from the steel industry show the relationship between good investments and market success clearly. But there are two critical elements to successful investment strategies. First, the individual investment decisions have to be made in light of a focused market strategy.

Second, the totality of a firm's investments have to fit together very well. A company can invest in technology, it can invest in the knowledge base of its employees, and it can invest in the way that work is organized and empower employees to make timely decisions. The payoffs to all of these investments are linked. Ignore one and the productivity of the other investments will suffer. Our study explains how firms can take advantage of this knowledge to prosper.

The large companies that restructured successfully and the most successful minimills align their management strategies and practices in the same direction so that they reinforce one another. Although the specifics varied depending on the individual company circumstances, in general top management set a clear vision of the company's future and communicated that vision very well. Trust was established, and responsibility was pushed downward throughout these organizations. Decision making was decentralized. All this was coupled with clear expectations for performance, and people were held accountable for the results over which they had control. In case after case, the combination of these characteristics helped establish the culture of the firm.

In a very real sense, the investment made in technology, in people, and in the way that work was organized set the stage for success. We will show that the culture of these firms helped to secure their success and sustain their competitive positions.

We began this work with independent research goals, but they were all related to restructuring. For one of us, management was the focal point. For another, it was technology and how industries can get the most out of it. For another, the market forces driving change had to be understood before much else could be accomplished. Slowly, each of us realized that substantial progress in our research was only possible if we worked together. Any one of these elements was woefully inadequate as a basis for explanation. Together, we had some hope of making the pieces fit.

This book is an outgrowth of that realization. It combines knowledge from economics, business management, and materials science, and the lessons involved have stood the challenge of criticism from each. We came to the end of our work with understanding that we could not have realized individually. We shared in the substance of the research, we shared in the writing, and we share in the results.

Pittsburgh, Pa. R. S. A.
November 1995 R. J. F.
F. G.

Acknowledgments

This book would not have been possible without the cooperation of more than a hundred individuals in scores of companies who filled out questionnaires, showed us around, answered our questions, and shared their insights. We are greatly indebted to them for their patience, openness, and willingness to commit vast amounts of time to us.

We would like to single out a few individuals who, more than others, helped to shape our understanding of the industry and the changes encountered over the past several decades. Since some of the people interviewed have subsequently changed position or company, they are identified by the company they worked for when we interviewed them.

Our thanks go to the following: Birmingham Steel Corporation (James Todd and Avery Hilton); British Steel, plc (Peter Ferguson, Edward Denham, and John Waine); Co-Steel Sheerness (John Clayton and Hugh Billot); Nippon Steel Corporation (Tomokatsu Kobayashi, Takahiro Hino, Toshihiko Ono, and Shigefumi Kawamoto); Nucor Corporation (John Correnti, Keith Busse, and Rodney Mott); Oregon Steel Mills, Inc. (Thomas Boklund, Vicki Tagliafico, Joseph Corvin, and Kevin Ratliff); Sumitomo Metal Industries, Ltd. (Chikashi Morimoto); Tokyo Steel Manufacturing Company (Masanari Iketani and Hitoshi Toita); and U.S. Steel Group of USX Corporation (Thomas Usher, Paul Wilhelm, Reuben Perin, Thomas Sterling, James Kutka, John Goodwin, John Kaloski, and John McClusky).

Special thanks go to Thomas Graham for his valuable insights and his extensive commitment to our project during the period 1990–1995 while he guided the U.S. Steel Group of USX Corporation, Washington Steel Corporation and AK Steel Corporation.

Ralph Gomory and Hirsh Cohen of the Alfred P. Sloan Foundation were prime movers of our project. The high goals and expectations that they helped to set with respect to the practical relevance of our work, its interdisciplinary nature, the involvement of our students, and academic excellence, helped bring focus to a very large undertaking. Over a six year period, the funding they helped to make available to us through the Foundation sustained the project and made it possible to pursue our inquiry on a broad front. Moreover, they were a sounding board for ideas and a stimulus for analysis. The American Iron and Steel Institute and the Nippon Steel Corporation also provided financial support in the initial phase of our work, and we would like to express our appreciation to them as well.

We are also indebted to many of our colleagues who participated with us in the study of the global steel industry, including Patricia Beeson, Richard Cyert, Anthony DeArdo, Gary Florkowski, Yuji Ijiri, Sok-Myon Kang, Carrie Leanna, Marvin Lieberman, Harry Paxton, Katherine Shaw, and doctoral candidates Tai-Hwa Chow and Craig Richmond. We are grateful for the support provided to various aspects of our work by the Center for Social and Urban Research at the University of Pittsburgh and its director, Vijai Singh, and the Department of Land Economy at the University of Cambridge.

Our work came to fruition because of the secretarial, editorial, and administrative support provided by Irene Petrovich at the University of Pittsburgh, Marilyn Paul at Portland State University, and Madeline Lesko at Carnegie Mellon University. We are deeply appreciative of their patience, good humor, and extensive assistance over a long period of time.

Finally, we thank Herbert Addison at Oxford University Press for his support, guidance, and encouragement throughout the final phase of this undertaking.

This book is published as part
of the Alfred P. Sloan Foundation program
on Centers for Study of Industry.

Contents

The Renaissance of American Steel

1

Introduction

In crisis there is sometimes opportunity, and so it has been for steelmakers in the United States. What was perceived by many as the collapse of this industry in the face of an onslaught by foreign competitors is now well recognized as the industry's rebirth. In this remarkable turnaround important elements of industrial competitiveness are exposed. They are known to some who have taken part in the renewal and to others as observers of the process, but more often the lessons involved have been buried by political rhetoric or by mounting social costs in dying steel towns.

Our intention is to lay a basis for understanding this dynamic that goes to the heart of the interplay between technology and management practices. We find that building and sustaining competitive advantage requires a continuous drive for improvement in both these dimensions, and we believe that this principle has broad applicability. Neglecting one or the other, and thus failing to recognize their complementarities, can make the crucial difference between success and failure, where the well-being of thousands is at stake and billions in profits can be won or lost in the competitive struggle.

But technology and management practices do not exist in a vacuum. The leadership provided by top management and the vision that flows from that leadership supply the framework in which investments in technology are made and decision making occurs. A company's competitiveness, therefore, is derived from the interplay of three factors: leadership, technology, and management practices, where management decision making embraces organizational issues as well.

Our analysis shows that successful companies have a conviction about where they want to go that is rooted in a deep understanding of products and markets. This vision is communicated broadly, and the strategies for carrying it out are well defined and clearly articulated. More, the strategies that they choose embrace elements of cost reduction, responsiveness to customers, quality, and productivity. They drive the company's investment in technology and in people. The decision making in successful companies is relatively fast, bureaucracy and organizational levels are minimized, and significant attention is paid to the motivation of employees through empowered decision making and financial incentive systems that reward performance. Basic to all this is building a culture of trust and commitment—one that focuses the goals of employees on the objective of striving for improvement on the job and in the production process.

The key to success lies in the interrelationships between these practices and strategies. There is no simple solution to building and maintaining a firm's competitive position. Whether the company is restructuring or is building itself anew, the principles summarized here and described in detail in later chapters form the basis for success. Of critical importance is the way in which these practices and strategies complement one another, for by their interplay a firm's competitive edge is honed.

Fundamentally, the decisions made by management to organize the workplace around principles like teamwork, pay-for-performance, or creating a well-trained workforce represent investments that may pay off or fail, just like investments in machines can pay off or fail. The "culture" of a firm flows, in part, from investments of this kind. It turns out that the value of such human resource practices for a particular firm can depend critically upon the ways in which all its investments fit together—investments in machines and in people. Our study demonstrates this. In view of this we stress a comprehensive, integrated approach to creating a competitive organization. This is not to say that "total quality management," "employee empowerment," or "business process reengineering" are not useful concepts or objectives, but they are not total solutions either. Rather, if implemented well they can be important elements in creating the conditions supportive of a successful company.

The conclusions drawn in this book are based on findings from a multiyear, multimillion dollar study of the steel industry supported by the Alfred P. Sloan Foundation as part of its global industry competitiveness studies, the American Iron and Steel Institute, and the Nippon Steel Company. The purpose of the study was to learn how U.S. steel companies could increase their global competitiveness. The study was organized into groups of scholars, twenty-six in all, who focused on technology, environment, government, and management issues.

We draw on a subset of the larger study that looked at the forces driving industry restructuring as a source of information on the basis for competitive success in steel markets. In this research, we were particularly interested in learning how the large steel companies such as United States Steel, AK Steel, British Steel, and Nippon Steel dealt with the need to reduce the size and scope of their operations to become more competitive, and having made the decision to reduce capacity, how they were able to reorganize themselves to reach more competitive thresholds. In addition we studied smaller steel companies to learn how high performers like Nucor, Birmingham Steel, and Oregon Steel in the United States, Co-Steel Sheerness in England, and Tokyo Steel in Japan have achieved their success. The objective of our investigation was to understand the policies and practices used by management to produce visible results in terms of high productivity levels and improved financial performance.

The loss of competitiveness in many U.S. industries was made explicit in the early 1980s, when foreign competitors seriously eroded the domestic market share of U.S. companies in industries such as steel, autos, computers, machine tools, and textiles. This set the stage for major restructuring in these industries, and the companies involved were "downsized," "repositioned," or "reorganized for success." These processes continue unabated to this day. But they are not limited to manufacturing industries. Rather the upheaval has broadened, and the ensuing change now touches industries as far afield from heavy manufacturing as banking, financial services, telecommunications, and wholesale and retail trade. Because of this we believe our insights are generalizable well beyond the steel industry. Whether one's responsibility runs to an entire company, a plant, or a department, those interested in improving the performance of their organization will find value in the elements of success we expose and analyze.

Dramatic Change

The change that has occurred in the steel industry is immense by any standard. Consider the 1980s alone. At the outset of that decade, over half a million American workers were employed in steel manufacturing.[1] Well over a quarter million jobs were lost in the decade to follow.[2] U.S. raw (unfinished) steel production declined by more than 12 percent, ending the decade at 98 million tons.[3] Enormous purges were made in steelmaking capacity. Plant closure and reorganization brought the industry's production capability down from 154 million tons in 1980 to 116 million tons by the decade's end—an adjustment of 25 percent. It is still lower today.[4]

Even these figures mask the devastation wrought in particular cities and steel communities. Pittsburgh, Youngstown, Bethlehem—these names were once synonymous with steel. In these places the identity of the people and the career paths of generations of families were inextricably bound to the well-being of particular steelworks, now and forever closed.

The United States was not alone in the crisis. Its consequences were felt by every steel-producing country in the world, no less by European nations than by those in the Americas. Even in Japan, a recognized leader of efficiency in steelmaking and a country at the vanguard of technological change in the industry, the extent of the adjustment has challenged traditional ideas concerning the lifetime tenure of jobs.[5]

The industrial organization in countries, the traditional role of governments in protecting steel producers and steelworkers, and the national cultures that determine the social safety net for workers and firms are vastly different around the world. But the crisis was widely shared. For example, in the face of depressed markets and limited prospects of recovery, the European Community invoked treaty provisions to manage markets and share the burden of decline, and provided massive subsidies to keep inefficient steel producers afloat. In the process they accomplished employment and capacity reductions without sacrificing political relations among member countries.[6] But the costs in terms of the competitiveness of European steelmakers in world markets may have been substantial.

Contrast the picture drawn by these realities with more recent comments of industry observers: "discussions with automotive industry executives indicated that for many automotive end-uses, domestically produced steel is as good as, or better than, Japanese steel in terms of quality."[7] "American steelmakers have so improved their productivity that the industry now often rivals—and in ways outperforms—its competitors in Germany and Japan."[8] "Who would have thought it? Entrepreneurs are racing into steel, not long ago one of America's sorriest industries."[9]

These gains come from two segments of the industry. The traditional firms that still dominate the popular image of steelmaking have shed unprofitable operations, reorganized production to eliminate waste, and focused their resources both in terms of product lines and of the geographic markets they serve. Industry change has been caused by international as well as domestic competitive pressure, and it is to the source of the domestic struggle that one must look for the second basis of the U.S. steel industry's new competitive edge. Independent producers, operating at much smaller scales than traditional steel firms, have entered the market and helped to transform it. These "minimills," as the independent producers have come to be called, pulled the industry to new levels of efficiency in technology and in the use of human resources.[10] And the competition they provided helped to propel productivity advances among the older, bigger companies.

The Basis for Competitive Gains

The turn of events in the U.S. steel industry has been chronicled by industry experts and the popular press.[11] These accounts often link cause and effect by focusing on external forces like unfair trade or changes in technology in order to enhance understanding of industry issues. Typically these explanations blame the bureaucracy of huge steel corporations or the shortsightedness of a monolithic steelworkers' union for the decline of traditional steel firms. They also view the newer independent producers as being uniformly aggressive, agile, and successful. But in our view, both images are lacking, and fail to capture the fact that much can be learned about competitive adjustment by drawing on the experience of the entire industry—traditional firms that weathered change and in the process achieved new competitive positions, and minimills whose inroads have redefined the industry's structure and its competitive dynamics.

Our work goes directly to the people and the firms involved in this struggle to expose the keys to competitive success. The building blocks necessary for learning involve principles of steelmaking technology and principles of management. The mortar in this understanding is the economics of the marketplace.

The importance of leadership by top executives to a company's long-term competitiveness is illustrated by the companies we have studied. For an existing company faced with crisis, there must be someone with the authority to make fundamental change, someone who understands when action is necessary and can translate that understanding into new success. The same kind of leadership and conviction are necessary for currently successful companies that want to grow and prosper further as a result of that growth. There are important junctures in the life of a firm when opportunity must be seized. Leadership is required to recognize those junctures and use the economic forces at hand to the firm's advantage. At those times a company's success may well turn on the vision of one individual to set a goal and guide the firm through necessary transformation.

Technology has often been a driving force at critical junctures for the steel industry. Due to the need to reduce cost and improve the quality of many steel products, since 1975 major new technologies have been developed and implemented. Initially, for a variety of reasons the big integrated producers were slower than their European and Japanese competitors and, in many cases, the U.S. minimills in implementing these improvements. The reasons for the delay in implementation included lack of profits and capital, not anticipating the changing product needs, and not recognizing the full potential of the technologies for cost reduction and quality.[12] Another reason is that with the existing labor agreements it was not always possible to fully exploit the benefits of technological improvements. The U.S. minimills, however, were often leaders in adapting technology. They made profits and could raise capital, had lower capital requirements and were able to take advantage of the productivity gains. Also, in general, the minimill management was closer to production and had a less conservative, more entrepreneurial culture.[13]

In the mid–1980s traditional integrated firms carefully invested in technologies that were required for their markets, often in joint ventures with Japanese partners, bringing about major cost reductions and quality improvement. In some cases they negotiated labor agreements to allow them to exploit these improvements more fully. Currently, the best U.S. integrated plants are on par with European and Japanese producers despite the fact that they may not have all the latest technologies.[14] They have been successful in making critical investments and doing more with less. The major U.S. minimills are the leaders in the implementation of the technologies associated with electric arc furnace production and set the world standard for efficiency in certain product lines.

Management practices have played an important part in bringing about the industry's newfound competitiveness. The rejuvenation of the large integrated plants has been accomplished in part because of greater market orientation and more attention to the involvement of the workforce in the decision-making process. Focusing more narrowly on higher value-added markets has enabled management to utilize limited capital expenditures to better serve these markets. Workers and managers at the plant level have been empowered to make decisions and to assume greater responsibility for their actions, and gains have been achieved in productivity, quality, and delivery.

While managers in a number of firms have found success in other ways, those in the top performing minimills seem to achieve success by trusting employees to make good decisions and giving them the support they need. The people that make up these firms are charged with decision-making authority in their sphere of influence, whether it is limited to the plant floor or extends to the whole firm. This gives rise to a thin organizational structure, decentralization, empowerment at all levels of the workplace, and ongoing communication. Moreover, supporting human resource policies, which include training as well as strong monetary incentives, are integral to the firms. These practices create a highly motivated workforce—one that accepts risk, is accepting of change, and strives for improvement.

The yardstick for success in market competition is simple. Ultimately the vitality of a company is reckoned in terms of its profits. Therefore one must look to elements that determine a firm's revenues and costs as the mechanism by which competitive advantage can be achieved and sustained.

The modern market for steel is wide open to competition, domestic and international. The implication of this for pricing is clear. The old days are gone in which a firm's market power translated nicely into the power to influence market prices. Now, if one firm enjoys a substantial margin above production costs, it will not be very long before others seek that margin too, unless a barrier to entry can be established that will restrain competition.

Successful firms take advantage of natural market barriers where they exist, and create barriers when they are not already present. We will draw on examples to demonstrate this and show that steelmakers have sought this competitive edge in technology, supply sources, and plant location itself. Some have sought it in government protection from foreign producers as well. What is a level playing field, after all?

Advantage on the cost side is both simple and complex. To be successful ultimately involves setting a very high goal: to pursue quality and efficiency simultaneously. The elements in achieving this goal, however, are many. Again technology is key, but this does not mean that one has to be on the cutting edge of technology to gain quality or cost advantage. Rather, it is the effective use of technology that matters, whether it is at the frontier or not, and management creates the environment and sets the incentives for this to happen. People make technology work, and they can make it work well or poorly.

In Chapter 2, we lay a basis for understanding the dynamics of change in this industry by describing the steel crisis and documenting the nature and extent of industry adjustment in response to that crisis. Because the technology of steelmaking is both a force of change and a source of continuing advantage in competition, our overview of the industry is complemented by a discussion of the machines that make and shape steel in Chapter 3. We use that chapter also to explain basic product lines in the industry and discuss the ways in which steel firms seek competitive advantage. We shall see that in important ways the future of the steel industry—the types of firms that will dominate it and their competitive standing in international markets—turns on the direction of technological developments.

Chapters 4 and 5 seek insight on the sources of competitive advantage by going directly to those engaged in the struggle for market dominance. The substance of these chapters draws on case studies undertaken for some of the most dynamic firms in the industry and on personal interviews with the people who have helped these firms to achieve leadership status. Lessons are taken from these examples of success, which span the types of firms that describe this industry. In these chapters we draw on extensive investigations of the firms and plants involved, observations of the people who operate them, and analysis of the market environment in which the success or failure of decisions is determined. Chapter 6 brings this information together in a way that exposes the complementary nature of technology and management practices in achieving and sustaining competitive advantage.

Our description and analysis of industry change are broadened in Chapters 7 and 8 where we turn to the experience of firm leaders in Europe and Japan for further insight on competitive adjustment. In these chapters, we also draw on case studies of integrated producers and minimills. Our objective here is to learn from the contrast and similarities between restructuring abroad and in the United States.

Chapter 9 examines the technological changes that are shaping competition in the steel industry today and will determine the competitive field of play tomorrow. From that backdrop, we use Chapter 10, the concluding chapter, to summarize our findings. Our objective in this chapter is to highlight the practical implications of our work for good management practice and its role in realizing success in global competition.

Summing Up

Our study shows that there is no silver bullet to improving competitiveness. There is no single, short-term solution for managers to grasp. While technology has played an important role in improving performance in the steel industry, the dynamics of success depend fundamentally on the people involved in production and on the leadership and vision provided by management. But their contribution cannot be elevated easily through a single, programmatic approach, such as linking monetary incentives to quality as *the* way to bring about long-term quality improvement. Superior results occur because of management's commitment to creating not only a work culture that motivates all parts of the organization to pull in the same direction but also to ensuring that all policies and practices support movement toward the company's goals and objectives. This requires a clear vision understood by all employees, and the linking of investments to what is required to compete successfully in chosen markets. It also requires the implementation of human resource policies to ensure that employees have the skills and the motivation to meet the needs of customers and to improve quality, productivity, and service. All this must be supported by an organizational structure that encourages flexibility, accountablity, and responsive decision making rather than one that impedes these critical elements.

The practices employed by managers in high-performing American, British, and Japanese steel companies illustrate many of the benefits that can be derived from concepts referred to in the management literature as "reengineering the corporation," "high-performing workplaces," "creating customer value," "total quality management," and "employee involvement," provided these approaches are implemented in a comprehensive manner. One of the interesting insights gleaned from our study, however, is that in many of the companies management did not set out with any of these programs in mind, and they eschewed narrow, programmatic views as they established a competitive position. Instead, their success was built on policies and practices that recognized that the desired results depended upon the interplay between all parts of the organization, from top management's communication of the goals and objectives to the types of investment decisions made, the types of markets served, and the incentives provided to improve performance.

We believe that the insights we have derived from studying how these steel companies have successfully improved their performance are broadly applicable. The examples provided should help managers at all levels, in the steel industry and beyond, to think through the ways in which they can operate differently to create the conditions necessary to sustain or enhance their competitiveness.

2

An Overview of Industry Change

The recession that followed the oil crisis in 1973 was a watershed for steel producers worldwide. By 1975 it was revealed to all that the world economy had substantial excess steelmaking capacity. The steady growth and sustained profitability that had characterized steel production in the previous decade gave way to uncertainty about the prospects for a rebound and pessimistic forecasts of future demand.[1] The reality of it was difficult to accept. For industrialized nations, the steel industry had been fundamental to their success. Employment levels in the industry assumed symbolic as well as political importance.

Indeed, throughout much of the world the state and the steel industry are one. In many nations, the industry is owned or at least nurtured by the government, which means that the extent and geographic distribution of worldwide steelmaking capacity are not determined by market forces alone. Steelmaking is emblematic of industrial development, and the governments of developing nations have often regarded the presence of a domestic steel industry as critical to their long-term goals. But because the technology employed in steelmaking often demands huge capital investments and requires very large rates of output in order to realize inherent economies of scale, smaller countries must depend on exports to achieve efficient production levels. Consequently, if demand falters and prices are depressed on international markets, there is strong incentive for the governments of these nations to provide the financial help needed to avoid plant closure or employment cutbacks.

Even in highly industrialized nations, the politics associated with employment in the steel industry can be intense. Plant location decisions have figured prominently in efforts to address regional economic disparities within European countries. Huge public investments in steel plants in southern Italy, in Wollonia, the French-speaking part of Belgium, and in Spain's Basque region are cases in point.[2] When these initiatives fail, the companies involved nevertheless serve as conduits of regional aid. When support from national governments is not forthcoming, local governments can take the lead either by exerting political pressure to help the industry or by direct financial support from local coffers. Here Germany is the prime example. That country can boast some of the best steelmaking facilities in the world, and the German steel industry is dominated by privately held firms. Even though the federal government is loath to provide subsidies to these firms, state governments within the federal structure of Germany are equally loath to see major employers fail.[3]

The result of this comingling of industry and government worldwide is shown dramatically in Figure 2.1. From 1970 through 1982 world steelmaking capacity ratcheted ever upward and bore little or no relationship to long-term trends in the worldwide demand for steel. By the mid–1980s, it was apparent that the gap between world capacity and world production had become unsustainable. Steel prices were driven downward, losses plunged steel firms into financial crisis, and plants were being closed.

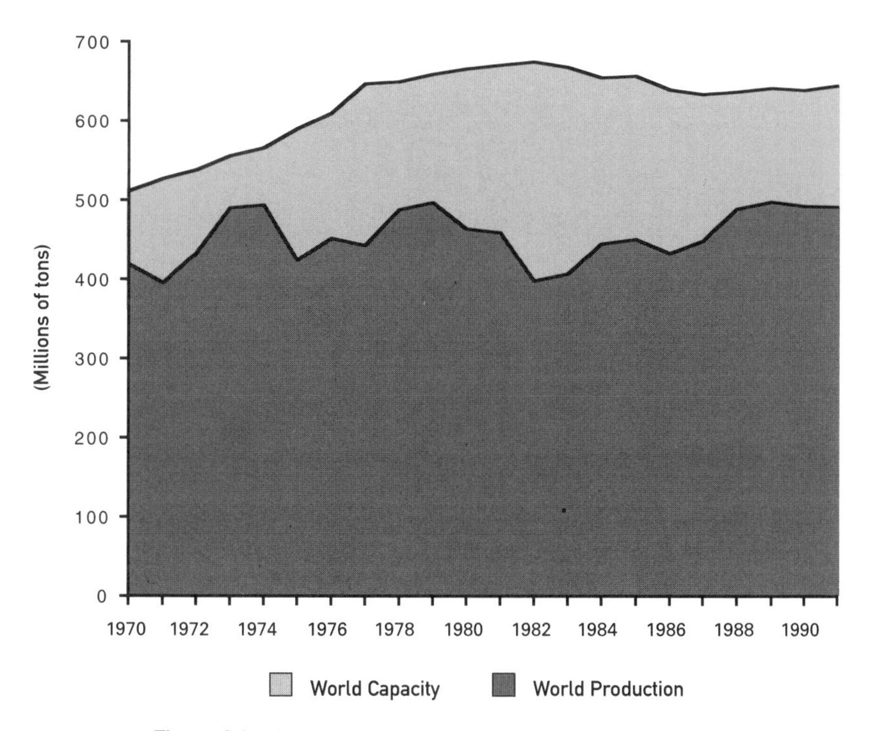

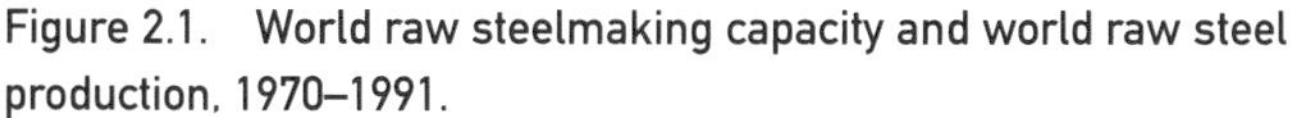

Figure 2.1. World raw steelmaking capacity and world raw steel production, 1970–1991.

Source: Paine Webber for world capacity and International Iron and Steel Institute for world production.

The American Steel Crisis

For the American steel industry, the seeds of competitive disadvantage were set in the aftermath of World War II. The mills in Pittsburgh and other steel centers were at the heart of our war machine. After the war, the debate centered on whether or not industry capacity should be expanded and at what rate. But there was little debate about whether the steelmaking technology in place during the war should be replaced. The giant open hearth furnaces that served us well through the war effort still had useful life, and would serve us well during the 1950s and 1960s too. At least, this was the gamble taken by industry leaders.[4]

A Technology Lag

Japanese and European competitors had to rebuild after the war. Only a small fraction of the furnaces they lost in the war's devastation were replaced by the traditional open hearth technology. The vast majority of their new capacity was in a new technology, the Basic Oxygen Furnace (BOF), which reduced the time it took to produce a heat of steel from over six hours in the open hearth to less than forty minutes.

Americans were slow to move in this direction. U.S. Steel and Bethlehem Steel built large open hearth shops in the 1950s when the new BOF technology was already available. BOF furnaces were also added in the 1950s and 1960s, but they were only a small part of our total capacity. Still more were added during the 1970s. By that time, however, global excess capacity in the industry had become a reality. The result is simple to describe: the lion's share of our capacity was in an old technology, while the capacity of our competitors was new and efficient by comparison. This American disadvantage in steelmaking technology was mirrored by a later disadvantage in steel casting technology. The Europeans and Japanese moved quickly into the world of "continuous casting," while we moved there only with a substantial lag.

In the 1970s the U.S. steel industry was dominated by firms that relied primarily on integrated production techniques in which iron ore is converted to crude "pig iron" and pig iron to steel in a highly-capital intensive process. Not only is the scale of individual plants typically very large in this subsector of the industry, but the firms are large as well. "Big Steel," the nickname sometimes attached to the integrated sector, resonates with the industry's popular image. In the prosperous times of the 1950s and 1960s these firms increased their capacity steadily and broadened their range of products. Production was up, demand was increasing, and their expectation for the future was bright: more of the same—lots more.

According to projections made at the time, raw steel production would be close to 200 million tons by 1994.[5] In reality, 1994 shipments of steel by U.S. producers were less than half that level.[6] There was a catastrophe in the making, which was deepened by the belief that there would be a massive increase in demand for iron ore pellets and an iron ore shortage as a result. From 1968 to 1980, Big Steel invested about $5 billion in new iron pellet programs and ownership positions in major ore mines. For every four dollars invested in the steel industry during this period, one dollar went to this purpose.[7] Moreover, the investments in ore and pellet plants continued well after it should have been clear that the earlier projections were flawed, and were crowding out investments that could have been vastly more effective in reestablishing global competitiveness.

Meanwhile, major ore development projects were under way in Brazil and Australia. The Japanese took full advantage of this by entering into long-term contracts that secured high-quality ore at very favorable prices. Indeed, the price of ore had dropped on international markets to the point where the Japanese could secure this critical resource at prices well below the cost incurred by American firms to get ore from domestic sources, even though the shipping costs for the Japanese were substantial.[8]

Failure in Labor Relations

If these investment decisions contributed to the crisis, the industry's labor practices worsened it immeasurably. The American steel industry is infamous for its adversarial labor relations. The carnage at the Homestead Works in Pittsburgh, when Henry Clay Frick unleashed his Pinkerton guards to attack striking workers in 1892, set the stage for labor-management struggles well into the next century.[9] In terms of the modern history of the industry, the strike of 1959 represents a critical juncture. It was long and costly. During that strike, the United States became a net importer of raw steel for the first time in the post–World War II years. We have imported more steel than we have exported by a wide margin ever since.[10]

At the time, industry leaders expressed the belief that strikes opened opportunities for foreign steelmakers in the U.S. market and that once those opportunities were seized, import gains were likely to be permanent. Based on this, the prosperity of the 1960s encouraged industry leaders to avoid costly strikes by buying labor peace.[11]

And did they ever buy it! The wages and benefits of steelworkers rose rapidly. Part of this story is revealed by the premiums that steelworkers have earned over the compensation paid to other manufacturing workers. In 1970 their hourly wage was some 27 percent greater than the manufacturing average. But by 1980 Big Steel's wage premium had soared to 67 percent.[12]

The drag on Big Steel's competitiveness caused by rising wages was made worse by its labor practices. The union had fought hard to preserve jobs and protect its workers from arbitrary dismissal by establishing work rules that governed how specific jobs would be completed and by whom. In inherently dangerous work, restrictions of this sort are natural, indeed necessary. But whatever the need, it is widely acknowledged that these protections went badly awry whether viewed from the perspective of management or labor. In far too many instances, productive time for workers in an eight-hour shift was minimal.[13] Of course, there are jobs in a mill that involve high risk of injury, and the threat of fatal accidents is ever present when the production process involves the transformation of molten steel. Under such circumstances rules to promote safety can easily justify regular rotations that provide for rest. But here we had a simple matter of redundancy on the job, redundancy that sapped the personal integrity of workers and built mistrust and contempt in management. Cooperation in the workplace was nonexistent.[14]

Ironically, labor's gains in the 1970s were secured just before a crash so large that its consequences would be felt for years. Figure 2.2 shows "apparent consumption" of steel in the United States, which is the sum of shipments by American producers to domestic customers and steel imports to the United States. From 1979, when domestic shipments by American steel producers peaked at just over 97 million tons, to the recessionary trough in 1982, apparent consumption of steel dropped by more than one-third. That loss has not been fully regained to this day.

Government Policy and Market Protection

One might reasonably conjecture that American steel producers would have shared in the prosperity that followed in the boom years of the mid–1980s, when the U.S. economy was creating new jobs at an astonishing rate. Instead, their crisis only deepened. U.S. monetary policy at that time was designed to purge the economy of inflation by maintaining very high interest rates. As a result foreign investors purchased dollars in order to buy dollar-denominated investments, which made the value of the dollar increase dramatically when measured in terms of other currencies.

This was terrific for Americans traveling abroad. A strong dollar means cheap meals and cheap hotels in London or Paris. It was devastating for U.S. steel producers. Buyers of steel in the United States could purchase foreign steel with the same strong dollars that made hotels and meals cheap for traveling Americans. Foreign steel producers were attracted to the American market as never before. As depicted in Figure 2.2, by 1984–85 they were supplying one out of every four tons of steel being purchased in this country.

These imports flowed into the United States to fill the demands of steel-consuming industries—such as autos, construction, appliance, and heavy

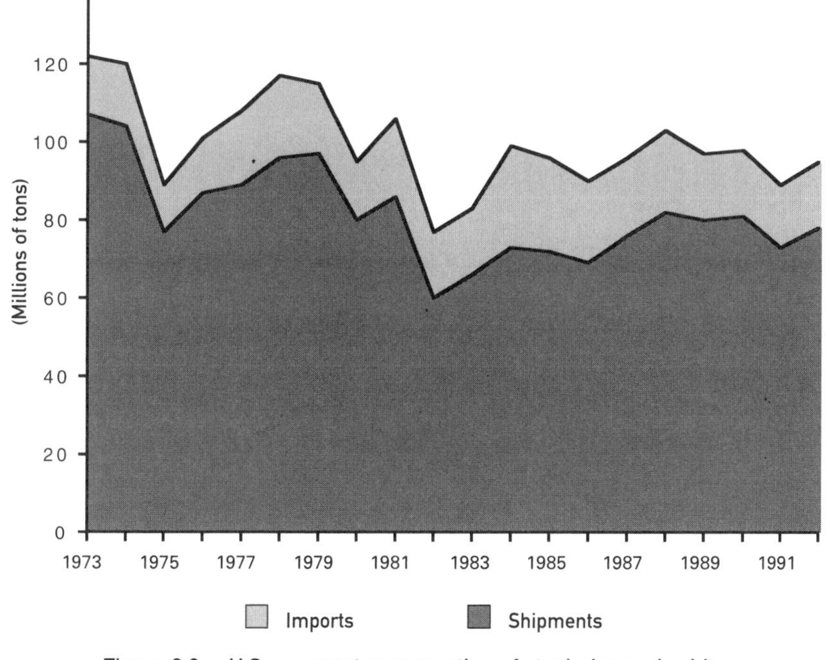

Figure 2.2. U.S. apparent consumption of steel: domestic shipments and imports, 1973–1992

Source: American Iron and Steel Institute. *Annual Statistical Report* (Washington, D.C.: American Iron and Steel Institute, various years).

equipment—for cheaper steel and, in some applications, for higher-quality steel. As early as 1982, import levels approached 20 percent of the U.S. market, and worse was yet to come. A disgruntled steel industry filed the largest number of trade cases in U.S. history. Altogether 132 cases were filed, 38 antidumping and 94 countervailing duty cases.[15] What followed was U.S. government action to diffuse industry pressure and compensate for its injury by restricting imports. Formal agreements were made between the United States and the European Community placing limits on Europe's exports. These "voluntary" restraint agreements (VRAs), first applied in October 1982, limited the market share of European producers to about 5 percent of U.S. consumption.[16]

Despite this agreement with Europe and despite Japan's independent efforts to voluntarily restrict steel exports to the United States, imports rose in 1983 and reached a high of 26 million tons (26 percent of total steel consumption) in 1984.[17] When Europe and Japan backed off, Spain, Korea, Brazil, Mexico, and Sweden, among others, filled the breach.

The rise in imports and depressed domestic demand reduced U.S. capacity utilization rates to below 50 percent in some cases and created large losses. More than 175,000 jobs were lost in the industry from 1980–84, and many plants closed.[18] Integrated producers piled up losses of $12 billion during the decade of the 1980s, which were offset by only $5 billion in realized profits.[19] Assets of $3.5 billion were written off as a result of closures and restructuring.[20] This gave rise to an intensive lobbying effort for stronger trade protection, with the industry advocating quotas.

To support this position, the industry filed dozens of new antidumping and countervailing duty suits, primarily against countries other than Japan and those in Europe. Yet more VRAs followed, this time with a string attached. Restraints were put in place for a five-year period, but they were subject to annual review.[21] Extensions would be granted year-by-year, and then only if the industry invested most of its net cash flow from steel operations in capital improvements and allocated about 1 percent of net cash flow for worker retraining. In return for reducing imports, the industry was required to modernize, and it did. Billions were invested in technologies that helped make the industry more efficient and improve the quality of American steel.

In the latter part of the 1980s U.S. imports fell by more than one-third,[22] and the industry did not want to yield its newly won protection from foreign competitors. As the time approached for the VRAs to expire the American Iron and Steel Institute, representing about 80 percent of the industry, advocated yet another five-year extension. They argued that this was essential in order to provide ample time for foreign countries to reduce their excess capacity, thereby alleviating their incentive to "dump" steel in U.S. markets. It also would give U.S. steelmakers sufficient time to continue their modernization programs.[23]

American steel users saw the situation in a somewhat different light. For them restrictions on imports translated to higher material costs and a diminished capacity to compete on world markets.[24] The debate was in progress at the time of the general election in the fall of 1988. As a candidate for the executive office, Vice President George Bush offered his support for an extension of the VRAs. In the next year, as President, Bush carried through with that pledge. The program was extended, and was coupled with an initiative to negotiate an international steel consensus agreement to remove trade-distorting practices.[25]

In March 1992, President Bush allowed the VRA program to expire. One of the primary arguments that had been advanced for its continuation was that the proposed international steel agreement (Multilateral Agreement on Steel Trade Liberalization or MSA) had not been negotiated.

However, imports into the United States were below the VRA limits, so pleas to continue VRAs until the MSA was in place were not very persuasive.

When the VRAs finally did elapse, the steel companies were free to file trade cases, and in June 1992 a dozen companies filed eighty-four dumping and countervailing duty suits against foreign steel companies in twenty-one countries. The suits alleged dumping and subsidy margins over 100 percent in many cases. Reaction by foreign governments and foreign steel companies was swift and highly negative. They saw the suits as commercial blackmail, a strategy to push up domestic prices and force agreement on the type of multilateral steel agreement advocated by the U.S. producers.[26]

The U.S. Department of Commerce had 115 days to make a preliminary ruling on the antidumping cases and forty days for the countervailing duty cases. The Commerce Department investigation found massive dumping and subsidy. In November 1992 the department imposed preliminary tariffs averaging 12.5 percent on the countervailing duty cases and in January 1993 it levied a preliminary tariff averaging 27 percent on the dumping cases, with some tariffs exceeding 100 percent.[27]

Half of these tariffs were later rescinded as a result of decisions taken by the U.S. International Trade Commission, and the American market was again wide open to foreign steel. By this time, however, foreign producers found that American steel was fully competitive in terms of price and quality. Restructuring by U.S. producers had gone a long way toward leveling the playing field in the international steel trade.[28]

The Steel Crisis in Europe and European Politics

The basis for today's European Union was laid in the Treaty of Paris in 1951, which established the European Coal and Steel Community. The intention of this treaty was clearly to minimize possibilities for conflict among member countries by binding together the interests of all the major steel producing nations in Europe. Subsequent treaties broadened the scope of cooperation greatly and led to enormous gains in terms of economic integration. It should not be surprising, therefore, that the European response to crisis in the steel industry is governed by the international laws that established this integration and help to maintain it.[29]

The Treaty of Paris created the institutional framework for coordinating the activities of all the steel companies within the Common Market, and it also coordinated public policy toward the steel industry in the member countries. The intent was to ensure that a stable, competitive market existed and that neither companies nor countries would take actions to give themselves an unfair competitive advantage. In the event steel markets destabilized, the structure provided the mechanisms to confront the difficulties. The decision-making process, however, was not simple or quick; it was deliberative and required significant consensus building. The interests of the individual states, the companies, and the workers involved had to be taken into account before action could be taken. This would necessarily require compromises along the way.

In general the European Commission, the governing body for the European Union (EU), as the alliance is now known, did not take an active interventionist role with respect to the steel industry in the decades up to the mid–1970s. Demand for steel grew throughout this period, and except for cyclical swings in the economy, European steel companies generally operated at high capacity utilization rates, and industry profits were substantial.

In 1975, however, in response to a decline in demand, European steel production fell by about 20 percent or 30 million tons, and it has never returned to pre–1975 levels. For the last half of the 1970s and for most of the 1980s, the industry operated at relatively low operating rates, a catastrophic condition for a capital-intensive industry.[30] This resulted in large operating losses for many companies, requiring significant government support to keep plants open.[31] Employment fell by more than 100,000 from 1973 to 1979 and by an additional 150,000 through 1982.[32]

Government Intervention

All this created a political crisis for European governments. The plant closures and large layoffs were at odds with European social objectives, and the immediate negative political fallout as unemployment rose provided strong incentive for state action. The situation was further complicated by the fact that the large steel companies in many of the countries were either owned, controlled by, or closely intertwined with government. So it was inevitable that industry problems would became central to the public policy agenda.

In the face of industry pressure to take even more drastic action the Commission instituted a program of "recommended" production cutbacks in 1977, a vain attempt to shore up prices by limiting supply. The voluntary nature of the plan proved its undoing. Many of the companies disregarded their quotas, and extensive price discounting occurred, followed by large industry losses.[33]

In order to provide a mechanism to better coordinate individual production decisions, the companies and the national steel associations in the EU member countries formed, with the encouragement of the Commission, a European cartel called Eurofer (European Economic Confederation of Iron and Steel Industries). The cartel began operations in 1977, and tried to establish production levels and set minimum prices for steel products in the member countries.[34]

Despite some success, however, price-cutting did not end. So in mid–1977 the Commission established mandatory minimum prices for certain products and recommended minimum prices for others. To ensure that increased imports did not undermine this plan, voluntary restraint agreements were established with the major countries exporting steel to the European market, and import quotas were established in some instances.[35]

These measures worked reasonably well in the later part of the 1970s, when demand strengthened and production levels increased. However, with the recession in 1980 and the collapse of demand for steel, the voluntary system fell apart. Capacity utilization rates fell below 60 percent, and unity within Eurofer dissolved as individual companies attempted to maintain volume through extensive price-cutting.[36]

A Managed Market

In late 1980, a condition of "Manifest Crisis" was declared by the Commission, brought on by the deteriorating market conditions and the large losses posted by steel companies. With the force of law, mandatory production quotas were set and previously established pricing rules were continued. The objective of the Commission was to provide some respite for the industry to enable it to strengthen its financial position and generate the resources required to carry out the necessary restructuring.[37]

The restructuring was difficult to achieve on a voluntary basis. Each country and each company wanted to maintain as much capacity as possible and consensus could not be achieved through Eurofer. The Commission estimated that there was more than 55 million tons excess capacity in 1980 and that capacity reductions of that magnitude were necessary to stabilize the market at current and forecasted levels of demand.[38] To achieve these reductions, plants and mills would have to be closed, and the greatest burden would have to fall on those companies suffering the greatest losses, receiving the largest subsidies, and running the most obsolete plants. This was a sensitive political issue, especially when each country began to realize how it would be affected. Eventually, the Commission agreed to cut annual capacity by 30 to 35 million tons, and required each member country to prepare a capacity reduction plan for the Commission's review and approval.[39]

Subsidies in whatever form by national governments to individual steel companies were prohibited by the Treaty of Paris. However, the Commission did not vigorously enforce that provision of the Treaty prior to the mid–1970s. Profitable companies do not need aid, and for the most part Europe's steelmakers were profitable. However, after 1975 state aid was commonplace and obvious, and was increasing in amount. Member countries tried to stave off plant closures and workforce reductions through massive subsidies, which were estimated to be about $20 billion in the period 1975–80 alone.[40] Conflicts inevitably arose between privately held European steel companies and those receiving government aid, which meant more political conflict.

The European Commission tried to bring order to this chaos by legitimizing subsidies on the condition that aid had to be tied to an acceptable capacity reduction plan. This link between state subsidy and industry restructuring, which began in 1981, continued in one form or another for more than a decade. It resulted in capacity reductions of about 31 million tons for Europe as a whole.[41] While significant, this still left Europe with 20 to 25 million tons of excess capacity by official estimates, and utilization rates were still unacceptably low.[42]

The financial cost for aiding the steel industry was high. The Commission approved payments of various forms of aid from 1980 to 1985 in the amount of $37 billion. Loans and grants accounted for about 75 percent of this, and most of the loans were subsequently forgiven.[43] In the European view all this was necessary to avoid protectionism and social chaos. The Europeans believed that if left to unfettered market forces the problems of this industry might well have undermined Europe's unity.

The crisis faced by U.S. steel producers was exacerbated by the European actions. At the same time that Eurofer moved to limit access to the European market, European steel producers were aggressively seeking new access to the U.S. steel market in order to keep their production levels up and limit their employment losses. As we saw earlier, the Americans cried foul when their losses piled up.

Japanese Competitive Advantage

The Japanese miracle, as some have described it, has been rooted in the ability to achieve competitiveness in industries for which Japan did not have a natural competitive advantage. The steel industry is a case in point—perhaps *the* case in point. Although Japan did not have the indigenous national resource base to support the steel industry, it was able to turn this disadvantage into an advantage by creatively sourcing raw materials, thereby achieving a competitive edge. Cooperation between government and industry was a critical factor in this success.[44]

In rebuilding its war-shattered economy, in the late 1940s and early 1950s the Japanese government set out to nurture the steel industry as the foundation on which other manufacturing industries could be built. The Ministry of International Trade and Investment (MITI) played a significant role in the early years of the development of the industry by coordinating expansion and modernization. Steel firms were required to submit their expansion plans to the government, and MITI helped ensure that the firms could borrow sufficient funds to carry out the plans. Priority for the loans was given to the firms that seemed best able to modernize, reinforcing the need to invest in new technology and equipment. The government persuaded private banks to make preferential loans to the steel industry at low interest rates and rearranged payment schedules for previous government loans. The government also gave favorable tax treatment to the steel industry, including accelerated depreciation rates, lower property taxes, and exemption from duty on imported machinery and equipment.

As Japan achieved self-sufficiency in steel production and began to export steel in the late 1950s, the emphasis turned to expansion in order to increase the scale of production, lower costs, and raise the quality. This required a coordinated partnership between government and the industry inasmuch as the critical raw materials had to be sourced globally. Long-term raw material contracts were negotiated by the industry, facilitated in part by government financing of the development of new mines abroad. Giant ships were required to transport the raw materials in sufficiently large quantities to reduce transportation costs. Construction of such carriers was arranged with the three-way cooperation of the steel industry, the shipping industry, and the government. In addition, construction of new steelworks was planned almost exclusively at seaside locations, and arrangements were made to improve port and harbor facilities with the cooperation of the central and local governments.

This coordination of policies eventually gave the Japanese industry a competitive advantage in its raw material costs. The result was a global powerhouse of steel production. Japanese steel was of the highest quality, and it was being produced efficiently with the best technology in the world.

In the late 1970s and early 1980s, however, decreased domestic demand in Japan and declining demand worldwide forced the Japanese industry to retrench. No country was immune to the problems created by the global capacity glut. The Japanese production system, which was based on a strong export base, became particularly vulnerable to U.S. efforts to restrict steel imports. Under intense political pressure, Japan limited its penetration of the U.S. market. Suffice it to say that the zenith of the Japanese steel industry occurred in the 1970s, and since that time the industry, like most of its foreign competitors, has had to cope with excess capacity and redundant workers.

Capacity Change in the U. S. Steel Industry

Against this reality of massive foreign subsidy and political contest over trade, one can look back at the recoveries from the recessionary troughs of 1975 and 1982 and see one extremely important element in common for American steel producers. Although the causes underlying these recessions differ markedly, a second look at Figure 2.2 reveals that much of the gain in steel consumption during each upswing was realized by foreign steel. American shipments increased in both recovery periods, but both times imports of steel increased at an even faster rate. American steel producers were counting on substantial revenue growth during these cyclical upswings, but their hopes were frustrated.

Wage costs were rising, production was falling, markets were being lost to imported steel, and all the while Big Steel was being challenged by new competitors on the home front. By the mid–1970s a few minimills learned that they could gain advantage over Big Steel by producing relatively simple products such as reinforcement bars in very small-scale plants. In order to do this, they relied on electric furnaces to melt scrap metal and continuous casters to shape it.

The advantage of these producers was enhanced by very favorable scrap metal prices in the United States. Open hearth furnaces can use 40 percent or more scrap in their charge. As this technology was replaced by the integrated producers with the more modern BOF capacity, the demand for scrap metal fell significantly, and scrap prices dropped. For electric furnace producers this translated quickly into a competitive advantage.[45] The price of their key input was falling rapidly and their production costs fell in tandem. When combined with advantage in capital costs, new flexibility in plant location, and a largely nonunionized workforce, the small mills began to hack away at Big Steel's traditional markets because of their cost advantage.

It is easy to predict that this combination of intense foreign and domestic competition would have a devastating effect on employment levels in the integrated sector of the American steel market and change the profile of the industry as a whole. The job count for the industry today, 238,000 workers, is less than 40 percent of that observed in the early 1970s,[46] and for the integrated producers average employment levels have fallen by more than 75 percent over the same period.[47] Hundreds of thousands of people lost jobs permanently as plants were closed or reorganized.

The change that has occurred in reaction to the economic and political forces we have described is enormous. While American firms in 1993 shipped only about as much steel as they did in the mid–1970s,[48] they did so in plants well equipped to operate efficiently in a smaller market, and their employees are among the most productive steelworkers in the world.[49]

In the aggregate, the nature of this change is well captured by looking at how production capacity has been redistributed among industry subsectors. For this purpose, it is useful to distinguish four groups of producers in the industry. In addition to the ore-based "Integrated" sector and scrap-based "Minimill" sector, there are two other groups that help to complete the picture. Independent electric furnace producers have been a small part of the U.S. steel industry for many years, but some of these plants did not take advantage of continuous casting technology. Rather, like many integrated producers, these independents cast molten steel using ingots and later transformed the ingots into semifinished shapes for further processing. For this reason it is important to distinguish this subsector from minimills that used more modern technology. We will refer to producers of this type as "Other Scrap-Based Mills" owned by nonintegrated firms. The last subsector of the industry is comprised of "Specialty" steel producers. These firms generally produce steel with high alloy content, most often stainless steels. They are defined by the product market they serve, and because of this, they have not been in direct competition with the other industry subsectors.

Table 2.1 shows the capacity change that occurred in each of these subsectors from 1974 to 1994, an interval spanning the most dramatic change. At the beginning of this period, the United States had a total steelmaking capacity of 185 million tons, but by 1994 its capacity had been reduced by more than one-third.[50] Integrated producers absorbed the bulk of this loss. Their capacity halved from 158 million tons in 1974 to only 77 million tons in 1994. In sharp contrast, the Minimill sector increased its capacity by 29 million tons during this period. Part of this increase occurred simply as a result of the reclassification of Other Scrap-Based

Table 2.1. Steelmaking capacity in the United States by industry subsector: 1974 and 1994

Industry subsector	Capacity (millions of tons)		Percentage change
	1974	1994	
Integrated firms	158	77	−51.3
Minimills	8	37	362.5
Other scrap-based mills	7	1	−85.7
Specialty firms	12	8	−33.3
Total	185	123	−33.5

Source: Steel Industry Data Bank, University of Pittsburgh.

Mills that had added continuous casters to their production line. Sometimes this came after the shutdown and repurchase of these mills. As a result, this type of capacity has nearly been eliminated in the United States, and the gains we show for minimills reflect this reclassification. However, most of the capacity gain by the Minimill sector—at least 23 million tons—was new capacity in the midst of a widely acknowledged capacity glut on the world steel market. This gain was directly at the expense of integrated producers. Restructuring in the Specialty sector was significant, but hardly of the magnitude of the integrated firms. Specialty producers shed 4 million tons of capacity from 1974 to 1994.[51]

Measures of Gain

As capacity shares shifted among these sectors and overall capacity declined, the steel industry was reshaped and reformed. What has been achieved in the process is truly remarkable. One of the most visible measures of success has been labor's productivity. This is especially important because it combines the effects of several related factors. Difficult and costly decisions were taken in the industry that resulted in the closure of many plants with outdated technology and inefficient production layouts. At the same time, strategic investments in new technology were being made to improve market positions, and human resource practices were being revamped in many firms to make them more efficient. These interrelated factors have driven the productivity of steelworkers upward. According to the U.S. Bureau of Labor Statistics output per employee hour in the steel industry more than doubled from 1982 to 1993,[52] placing it well ahead of the productivity gains shown by the manufacturing sector as a whole.

In the steel industry, finding a position in the market that allows you to make the most of the technology and people you employ means operating your machinery at high rates of utilization. The importance of this cannot be overstated. When billions of dollars of capital are tied up in plant and equipment, idle machines translate into huge losses. In the 1970s utilization rates in the industry often hovered between 73 to 78 percent. In 1982, the worst year of the industry's crisis, utilization fell below 50 percent. With the restructuring that followed, utilization rates have improved dramatically. In the last few years of the 1980s, utilization moved up to 85 to 90 percent, and as the recession of 1991–92 faded, utilization rates again returned to this high level.[53]

Surveys of steel users concerning product quality and service reflect similarly on industry gains as U.S. steel producers have shed the mantle of failure and established a new basis of customer confidence. Customer surveys conducted by the U.S. International Trade Commission confirm that American steel producers meet or exceed international industry standards in these dimensions too.[54]

The quality requirements of Japanese auto transplants in the United States were one of the driving forces behind this change. Now, American car manufacturers have added their demands for better quality and timely delivery. As compared to the mid–1980s rejection rates by automobile producers have decreased by over 90 percent,[55] and American steelmakers have earned numerous quality awards from the auto industry, foreign transplants and domestic producers alike.

The massive restructuring of the U.S. steel industry was accomplished largely without government intervention except in the area of international trade where VRAs held imports below their natural level for much of the 1980s. As a result, the restructuring process was carried out by the companies themselves, seeking to shed unprofitable, inefficient operations, and raise productivity, quality, and profitability. This brought about a quicker and more dramatic restructuring than occurred in Europe, where high government subsidies propped up inefficient producers and reduced the need to close unprofitable operations, or in Japan where a strong social compact existed between the companies and the workforce, which effectively precluded layoffs.

Summing Up

Differences in the competitive context of change in the United States, Europe, and Japan contributed to differences in the form and extent of restructuring as steel firms responded to the global crisis. In the United States minimill gains are an important part of the industry's rebirth, but they are only a part if it, not the whole. It would be wrong to look at the numbers that portray capacity adjustment in this industry and identify minimills as the winners and integrated firms as the losers. Markets are more complicated than that. The change that underlies the industry as we know it today has resulted in strong, highly competitive firms in both these sectors. Interestingly, winners of both types—minimills and integrated firms—have achieved success in very similar ways.

Minimills have, on the whole, made good strategic investments, and taken full advantage of existing technology to exploit market opportunities, but they have only rarely been technology leaders. Their success is based on choosing the right technology for the right market, organizing the workplace to take advantage of that technology, and motivating people to excel. In an important sense, the integrated producers who have secured a place in today's steel market have succeeded for exactly the same reasons.

Confronted with the stark reality that they could no longer compete in the full range of steel products, they were forced to choose markets where they believed it was possible to create and sustain a competitive advantage. They chose product lines and associated markets carefully, and used technology to help secure and maintain market position. As important, they too strived to create a work environment in which management and workers are motivated to succeed, and where success translates into profits for the firm.

In the following chapter we will lay a basis for understanding competitiveness in this industry by describing the technology of steelmaking and the nature of steel products. In the process we will expose critical elements of this industry's success. Then we will return to these elements in later chapters to explain how competitive advantage has been established and sustained in modern steel markets.

3

Elements of Competitiveness

The realignment of steel markets in the United States has been driven by the decisions of leaders in a large number of individual firms. Some of the decisions were bold and involved major risks; others were taken at the margin with little to be lost or gained. Some were in response to perceived opportunity; others were purely defensive, a reaction to realized losses and a recognition that traditional battle lines had been redrawn.

As our analysis develops, we will find that the first and most critical challenge in the path to competitiveness is knowing the market that one wants to serve and understanding how to secure it. In this chapter we explain the nature of steel markets, and examine some of the economic drivers that have shaped the realignment of these markets among industry subsectors. We shall find that technology shapes not only competitive advantage but also the way that work can be organized to sustain that advantage.

The process of adjustment in this industry is rich with examples of the ways in which firms pursue market position. These examples are revealing for several reasons. Their aggregate effects were enormous. The changes occurred in a relatively short period of time and engaged an entire industry. They involved markets that were defined in two dimensions, by geography and by product line. The decisions were strategic and had the objective of gaining access to new markets or redefining established market positions to improve or reestablish competitive advantage. And while the role of the U.S. government in setting policy for fair international trade affected the extent of change and to some extent the type of change, major investment decisions were largely unfettered by local or national politics, including decisions about plant closure.

A firm is competitive if it can establish and maintain a market position in a way that generates profits. The name of the game is to know the price structure of available product lines, and pick one where your firm can achieve a cost advantage over existing and potential rivals. However, you have to draw the battle line and pick a market segment knowing that even if you can beat existing producers by widening the gap between price and costs, other producers can learn by your example and seek the same margin in the same way. Establishing an advantage and keeping it are not entirely the same thing.

Market entry always involves risk or it would not be worth the effort. Steel markets are highly competitive, in part because basic steel products are not readily distinguishable once they leave the factory gate. In some markets the quality of product is very uniform; in others quality matters, but there is no unique formula for achieving it. With effort, financial resources, and time anyone can get it right. Steel is a commodity, plain and simple. On the demand side its characteristics can be specified clearly, and steel with those characteristics can be obtained from many suppliers, domestic and foreign, at prices constrained by world markets.

Understanding Basic Product Lines

So what are the markets for steel? Steel is the quintessential material input. People want it (that is, firms want it) because they can use it to produce other things. The markets for steel break down by the industries that need it, either to help produce other goods or as a building material. It is sometimes easiest, and quite useful, to describe steel markets by the shape of the steel products that come out of a mill. These are either "long" or "flat."

Long products include rods, bars, wire, and angles—one can envision long strands of molten steel solidifying into the crude forms from which these products could be cut or shaped. Indeed, the reinforcing rods that are used in concrete, the bars that comprise our fences or gates, and the angles that help to support our roofs begin their existence indistinguishably from the same basic shape, a semifinished strand of steel about six to eight inches square or "billet." Beams used in construction come from larger semifinished shapes, often with cross sections measured to ten or as much as thirty inches, called "blooms."

Flat steel products go into cars and appliances most frequently, but the sheets of steel that form the surfaces of these products begin as enormous "slabs," typically five feet wide and eight to twelve inches thick. The same slabs can be rolled into plates rather than sheets in order to provide for the sides of ships or shaped into pipes with seams. Recently, technology in the industry has advanced and much thinner slab castings are now possible.

From these simple descriptions, the end markets for steel emerge. The construction industry alone accounts for about one-third of all steel used in the United States. The second most important steel-using industry is automobile manufacturing, which accounts for up to one-quarter of the domestic steel market. The rest of the market is comprised of machinery and equipment producers, appliance manufacturers, and the oil and gas industry.[1]

Within each major product category, long or flat, one can define products of higher or lower "value added"; that is, the price of these products may reflect a greater or lower markup over the simple costs of purchased energy, materials, and services. In the bar market, for example, the low end is defined by reinforcing bars or "rebars" and the high end by special bar quality (SBQ) products, where strength or elasticity may be crucial. In flat products, the metallurgical properties of the steel, its surface quality, or the coatings applied help to define the value-added range.

Risk in Choosing a Market

The risk inherent in steel markets is reflected clearly in price fluctuations. Consider, for example, the history of prices for hot-rolled sheet steel (a staple of the auto industry) as depicted in Figure 3.1. Producers of sheet steel enjoyed a steady increase in price through the early 1980s, but prices peaked in 1984. After that the roller-coaster ride began, with prices plummeting twice in the next nine years. By 1993, the average price of this

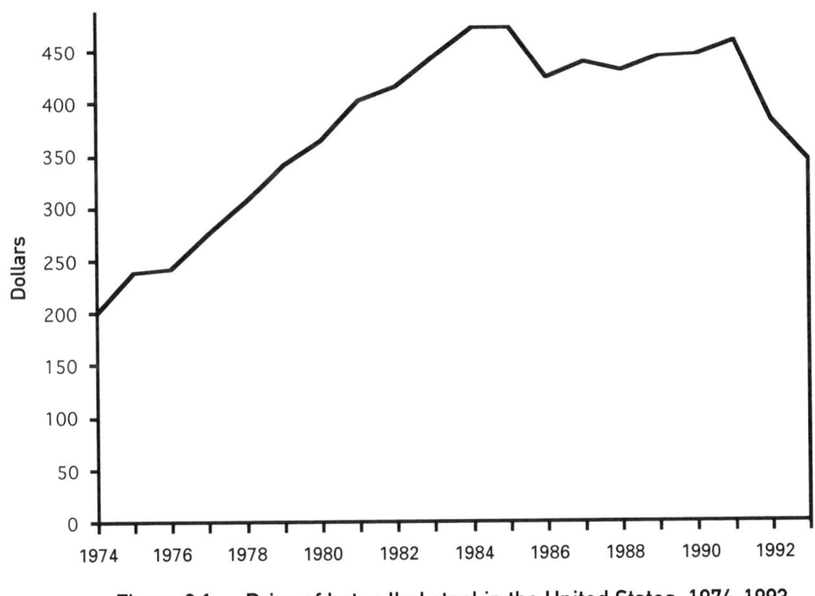

Figure 3.1. Price of hot-rolled steel in the United States, 1974–1993
Source: American Metal Market. *Metal Statistics* (New York: American Metal Market, various years).

product was only marginally higher than it had been in 1979. Needless to say, the cost of the inputs needed to produce steel did not fall in tandem with the price of steel over this period, and the profitability of many steel firms reflected this. The red ink flowed.

Intense competition in the U.S. market for steel explains the fluctuation in prices. During the 1980s the primary source of competition in this market was foreign imports. As explained in Chapter 2, imported steel flooded America in 1984 and 1985, when foreign producers accounted for more than one-fourth of all sales in this country—the highest level of import penetration ever recorded here. Their success in the U.S. market drove prices downward as American steelmakers struggled to maintain market share.

In 1992, domestic steel producers were recovering from a recession, and steel production was up over that of the preceding year by almost 6 percent.[2] The average market share of imported steel products had hovered in the neighborhood of 18 percent for four years .[3] In flat-rolled products, the market share of imports was up over that of 1991, but only by a small amount, yet prices plunged again.[4] Why?

This time the answer lies in intense competition among domestic steelmakers. New, low-cost production had come on-line for hot rolled sheet steel, and a new U.S. producer, Nucor, was claiming market share long reserved for traditional firms. Nucor won over buyers with lower prices and reliable delivery. Today, this firm and a number of other U.S. minimills are looking for an even larger share of that market.

Restraining the Competition

In the end, it makes little difference to a displaced factory worker or manager if the job that was lost somehow found its way overseas or to another region of the United States. Similarly for those who stand to reap the potential profits that come with ownership, the blame for lost income due to business failure is often difficult to place, but the losses are real nevertheless.

Risks associated with competition can be limited, and sometimes the easiest mark is the foreign devil. In the U.S. steel industry domestic firms have often sought protection from imports by recourse to trade laws that are designed to limit unfair trading practices. As we have seen, their justification for seeking this protection can have a substantial basis in fact, although this is not always the case. For many reasons, some economic and some political, steel producers in other countries have received enormous amounts of financial support from their governments.

Now, however, the Europeans find themselves sharply divided over steel subsidies. Privately held firms in the United Kingdom and Germany cry out for protection from steelmakers that are kept from the auction block only by government largesse, and these pleas echo the logic and tone of American steelmakers who seek protection of the same kind.[5]

The U.S. economy is among the most open to foreign trade in the world, and because of this it is a rare American firm that can rely for very long on trade laws to limit the scope of the market subject to competition from foreign producers. In a highly competitive environment, one finds long-term protection in the quality and service provided to customers and, most importantly, in one's ability to maintain a cost advantage that prohibits a serious challenge to one's markets because the profits are simply not there for other producers, foreign or domestic.

In an industry where transportation costs are large, an effective market boundary sometimes can be defined in geographic terms. That is, a cost advantage can be established by access advantage to markets or raw materials. If your plant is close to prospective customers and you can get your product to them more cheaply than other firms, the profit margins of competitors can be limited. They will have to find advantage elsewhere—in production costs, for example—to extract profits from sales to customers near you. Access to raw materials can result in the same kind of advantage. A firm that can secure raw material supplies more efficiently than other firms and sustain that advantage, can squeeze more profit out of every sale. Effective geographic boundaries on a market can be set by the cost advantage earned on the supply side in this way.

Transportation costs matter in the steel industry for international as well as domestic shipments. A degree of protection from foreign steelmakers is afforded by the shipping costs they must bear in order to reach the U.S. market. So by their very nature, plants in coastal locations are more open to market penetration from imports than plants in the interior of the country. Similarly, domestic firms can take advantage of shipment costs within the United States to stake out regional markets, and historically some firms have done just that to gain market advantage.

Technological Advantage and Steelmaking Processes

In steel, truly important cost or quality differences can virtually always be traced to technology. The way steel is made—the choice of furnace and raw materials, the chemistry of the metal, the way it is cast, the way it is finished—lays a basis for defining a firm's competitive advantage. However, the importance of technology goes well beyond its engineering aspects.

Technology shapes the workplace. It helps to define the most efficient size of individual plants and by extension the most efficient size for the firms themselves. In turn, this can establish a framework by which management is structured and information about production or marketing is exchanged within an organization. The flexibility of the technology can affect the ability of the firm to respond to customer needs, or its ability to take advantage of price changes for key inputs so as to minimize costs. To understand these facts and their importance in the restructuring of the U.S. steel industry, it is essential to know the fundamentals of steelmaking processes and the basis for cost advantage that they help to establish.

Ore-based or integrated production and scrap-based electric furnace production are the defining methods of manufacturing steel. The competitive advantages and limitations imposed by these distinct processes shape the structure of the steel industry as well as the competitive margins on which specific markets are won or lost, in terms of product lines or geographic areas.

Ore-Based Production

Integrated steelworks involve a complex series of individual processes that use coal as the primary energy source and iron ore as the basic raw material. The crude ore must be concentrated by eliminating most of the unwanted materials, and then be shaped into "sinter" or "pellets." Coal must also be prepared for use in the steel mill. It is transformed into a derivative called "coke" in order to increase its energy value and strength. Both these operations, ore to sinter or pellets and coal to coke, are often accomplished in plants that are directly associated with iron and steelmaking operations. They involve large capital costs, and are carried on at high cost to the environment. Coke production especially results in highly dangerous and noxious discharges of water and gases. The costs of bringing such plants into compliance with environmental laws can be staggering.

The ore pellets and/or sinter, coke, and lime or limestone are fed into the top of a "blast" furnace to make pig iron. The scale of this operation is immense. A state-of-the-art-blast furnace has an internal diameter of 14 meters (45 feet) and a volume of 2,000 cubic meters. They are capable of producing up to 10,000 tons of metal per day. Air, often enriched with oxygen, is injected under tremendous pressure (blasted) and combusts with coke, or coke and pulverized coal, to produce the energy required by the process.

The crude pig iron or "hot metal" that comes from the blast furnace contains impurities. Removing them involves several steps. Hot metal treatment and desulfurization take place first, and then the metal is moved to an oxygen steelmaking (OSM) vessel. There are a number of versions of these. The traditional OSM is a top blown basic oxygen furnace (BOF), which is designed to remove carbon and other impurities from pig iron. Other OSM vessels are designed to accomplish the same thing—to render steel from pig iron by removing its impurities—but they might be bottom blown or use a combination of top and bottom blowing.

A typical BOF furnace treats two hundred tons of metal in about forty minutes including charging, oxygen blowing, and tapping. When tapped or poured from the furnace, steel is drawn into a ladle where its final cleanliness and chemistry are achieved. For high-quality products such as automotive sheets, the steel goes through a vacuum degasser as part of this further refining. The liquid steel is then cast into slabs, followed by hot and cold rolling into sheet. Finally, sheet can be galvanized (coated with zinc) or tin plated (coated with tin) to achieve the desired surface quality.

It will prove useful in terms of understanding technology differences as a source of competitive advantage if we summarize this information on ore-based production by reference to standard cost components in the manufacturing process. Table 3.1 categorizes the costs associated with the production of pig iron and those associated with integrated steelmaking up to (but not including) the costs of casting and rolling the liquid steel into its first semifinished shape. The costs are represented as four basic components: capital, labor, energy, and materials.

For an integrated plant to be competitive it must take advantage of economies of scale. Typically, coke plants have a capacity of 1 million tons per year; the blast furnace used to make pig iron typically operates at a scale between 1.5 and 3.0 million tons per year. The capital necessary to build new facilities of this kind might easily run to a total of $0.8 to $1.5 billion. None have been built in the United States for many years, and the economics responsible for that are not going to change anytime soon. However, even to refurbish existing iron-making facilities completely, it might cost as much as $500 to $750 million.

Capital expenditures like these are reflected in two ways when reckoning the unit costs of production. First, a portion of the capital costs must be taken each year as a depreciation expense, which has to be spread over each ton of metal produced. Second, there is an "opportunity cost" associated with the capital expenditure. Once committed, the money used to buy plant facilities and furnaces cannot be used for other investments. This means that one has to forgo the income from alternative investments as long as that money is tied up in the steelworks. For this reason, economists count the "normal return" on such assets as a cost (an opportunity cost) of capital. Both the depreciation expense and the opportunity cost of capital are "fixed" in the sense that the firm bears these costs independent of the amount of steel that is actually produced in a given year. So if a million tons

Table 3.1 Production costs for one ton of pig iron and one ton of steel in an integrated facility with three million tons of steelmaking capacity

Input	Pig iron production	Steel production
Capital	$	$
Depreciation	19	15
Normal return	19	15
Subtotal	38	30
Labor	11	11
Energy		
Coke	39	—
Other	4	3
Subtotal	43	3
Materials		
Ore/pellets	53	—
Pig iron	—	132
Scrap metal	—	33
Other	10	29
Subtotal	63	194
Total cost per ton	155	238

Source: Estimated by Richard Fruehan.

are being produced the capital charges can be spread over all of them, but if a plant is idled by strike or recession, the capital charges do not diminish at all.

In the production of pig iron, these combined capital costs are estimated to be $38 per ton (depreciation of $19 per ton and return on investment of $19 per ton), assuming, of course, that the plants and furnaces are being used to their normal production capacity. This amounts to an effective depreciation rate of 8 percent per year and a return on capital of 8 percent a year, both common assumptions in such calculations for the industry.

As we have explained, the materials associated with pig iron production are primarily ore pellets or sinter, but material costs also include the cost of flux and other supplies that are needed. Similarly, while the energy here is primarily from coke, small amounts of coal and other energy sources are involved in the process as well. Very little direct labor is used, barely more than one-third hour of labor time per ton of hot metal. This is reflected in the small share of labor expenses in total costs. Taken together, the cost of all these "variable" inputs (labor, energy, and materials) adds up $117. This brings the total cost (fixed plus variable) incurred to produce one ton of pig iron to $155.

The breakdown of steelmaking costs is similar in structure although the materials and energy involved are vastly different. In integrated steelmaking, pig iron is the primary material used. Approximately 0.85 tons of pig iron are required to make a ton of steel in an efficient facility. This material combined with the other required materials such as scrap metal, oxygen gases, and alloys, account for nearly 82 percent of the total costs of a ton of steel: $194 of a total of $238.

Looked at in this way it is easy to explain the possible sources of cost savings in integrated steel production and, therefore, the ways in which firms seek advantage as low-cost producers. Let us examine labor costs first. When so little labor is used per ton of output (in pig iron or steel) it is hard to see at first glance that labor inputs could be an important source of advantage. However, as labor productivity is increased, consider the possible cost consequences. Labor costs in a unit of steel are calculated as LxW (the labor used per ton {L}, say 0.33 hours, multiplied by {W} the hourly wage). Greater productivity per worker means more tons produced by the unit of labor involved. So instead of using one-third hour of labor to make a ton of steel, a firm might need three-tenths (0.30) hours of labor or less. There is a small savings eked out here, but it could easily be eroded by compensating increases in labor's hourly wage (W). However, more output per worker over the course of the year also means that capital costs can be spread over more units of output, and per unit capital costs can be driven down too.

Scrap-Based Production

The electric arc furnace (EAF) process uses steel scrap as its primary raw material. It is much less complex than ore-based production. None of the capital costs associated with ore preparation or coke production are borne by scrap-based producers. They begin directly with steel as the primary material input (albeit scrap steel), and their task is simply to transform it into a more usable form.

Some of the scrap used in this process is generated right in the mill as a result of yield loss in previous heats of steel, and for this reason it is referred to as "home" scrap. However, most of the scrap steel comes from outside sources, whether it originates from wastage in steel-using industries (prompt scrap) or from discarded consumer products (obsolete scrap). Whatever the origin of the scrap charge, the purpose of these plants is simple: melt the scrap as quickly as possible, immediately refine it in a ladle, and cast it into billets, blooms, or slabs—all without the delay or cost that would follow if the melted scrap metal were allowed to solidify between these steps.

Lower value-added steels can be produced in the EAF furnace using only scrap, and even the quality of scrap is not limiting. For example, it does not matter if the scrap metal in the charge contains high residuals (e.g., other metals or oils) if you are producing reinforcing bars. Because the bars are going to be buried in concrete, the surface quality of the material is not really relevant. However, in other products these residuals might matter very much and the scrap used would have to be carefully chosen to avoid excessive impurities. In the production of slabs from electric furnace operations, not only is higher quality scrap demanded, but part of the material charged into the electric furnace must be iron without the impurities contained in scrap. This could be pig iron from a blast furnace; direct reduced iron ore (DRI), a further refinement of iron ore pellets; or Hot Briquetted Iron (HBI), iron ore fines that are reduced to the form of a "pillow" only one inch square.

There are no HBI plants and only one DRI plant in the United States. The cost and availability of natural gas in this country is the limiting factor in producing these products domestically. Because of this, almost all DRI/HBI used here is produced in other countries. However, as the importance of these refined forms of iron has grown, some U.S. companies have begun to take equity positions in off-shore DRI/HBI plants and others are building them or evaluating major investments in these critical inputs.[6]

Once again it will prove useful to summarize by examining the production costs. Table 3.2 presents costs of EAF steelmaking (not casting or rolling) for three broad product categories: (1) a product for which low-quality, low-price scrap would suffice; (2) one in which high-quality, high-price scrap is the only metal used in the charge; and (3) a product in which a blend of high-quality scrap, low-quality scrap, and DRI/HBI is used.

Notice first that the capital costs associated with each unit of scrap-based steel are much lower than those associated with ore-based steel. An electric arc furnace plant capable of producing 1 million tons per year would cost approximately $160 million dollars to build, not including the costs of the casting and finishing machines. Operating at normal utilization rates, this brings the per unit capital charges down to $26 per ton ($13 per ton for depreciation expense and $13 as opportunity costs), which is well below the charges in an ore-based plant. Again, we apply industry standards of 8 percent for depreciation and 8 percent for the opportunity costs of capital in calculating these capital charges.

Variable costs in scrap-based production are very sensitive to the cost of materials in the charge. Labor and energy costs vary little or not at all as

Table 3.2 Production costs for EAF steelmaking in a plant with one million tons of steelmaking capacity

Input	Products based on		
	Low price scrap ($110/t)	High price scrap ($150/t)	Blended Change
Capital	$	$	$
Depreciation	13	13	13
Normal return	13	13	13
Subtotal	26	26	26
Labor	10	10	10
Energy			
Electricity	20	20	23
Other	9	9	9
Subotal	29	29	32
Materials			
Scrap metal	121	165	104
DRI/HBI	—	—	48
Other	19	19	19
Subtotal	140	184	171
Total cost per ton	205	249	239

Source: Estimated by Richard Fruehan.

one moves from lower- to higher-quality steel products, but high-quality scrap or blends of scrap and DRI/HBI come at a premium.

These cost variations establish the basis by which steel markets can be contested. In markets where low-quality scrap can be used, EAF steelmaking has an obvious cost advantage over integrated steelmaking. Competitive advantage like this drove integrated manufacturers out of long products and into flat products, where quality matters greatly. Now, however, the integrated firms find that even this traditional market stronghold, the market in flat products, is being challenged as technology advances and the EAF producers gain experience in its implementation.

We shall see later that the battle lines in this contest are still being drawn. Firms in each sector are assessing risks in light of the information available to them and trying to stake out new markets or defend their position in existing ones. In the drive to succeed, perhaps the single most important decision that must be taken is that of focusing the objectives and energy of the firm on particular product lines. In the steel industry this decision is based on opportunities shaped by the state of technology as well as the vision of leaders who believe that new advantage can be established by the way in which technology is put to use.

Steel Plant Characteristics

To a considerable degree, the technology used to produce steel has influence that extends well beyond defining the feasible set of product lines. Technology shapes the size of steelmaking plants, their input mix, and labor structure. Indeed, because of this technology helps to shape the very structure of the corporation itself. Integrated plants are large, achieving necessary economies only at a scale of 3 to 4 million tons of raw steel per year, at least. The production process is complex and necessitates coordination among a number of facilities, some producing or processing raw materials, some engaged in iron and steel production, some engaged in rolling and shaping operations. In contrast, the scrap-based minimills operate efficiently at very low rates of output, say 0.5 to 1.0 million tons per year, and coordination problems are minor compared to those in traditional ore-based firms.

Consider the operations of the USX plants in the Pittsburgh region as a case in point. Three facilities are located there, an iron and steelmaking plant in Braddock, Pennsylvania (the Edgar Thomson plant), a coke facility in the town of Clairton, and a rolling facility (the Irvin plant) near Dravosburg. Geographically, one might think of the plants as being three points on a triangle laid out roughly as shown in Figure 3.2. The meandering Monongahela River is also illustrated in the figure. Together, the operations take their name, The Mon Valley Works, from their proximity to that river.

The coke produced at the Clairton plant is shipped about ten miles to the steel plant in Braddock. In turn, slabs from Braddock are shipped six miles to the Irvin plant where they are rolled into flat bands that are coiled for sale to appliance manufacturers and other end users. While this is an extreme case of the complexity involved in integrated steel production because coordination is required among plants that are spatially separated from one another, it drives home the fact that in ore-based production several distinct operations have to be carried on at once. The capacity of any one component can limit operations in the others. The technology and

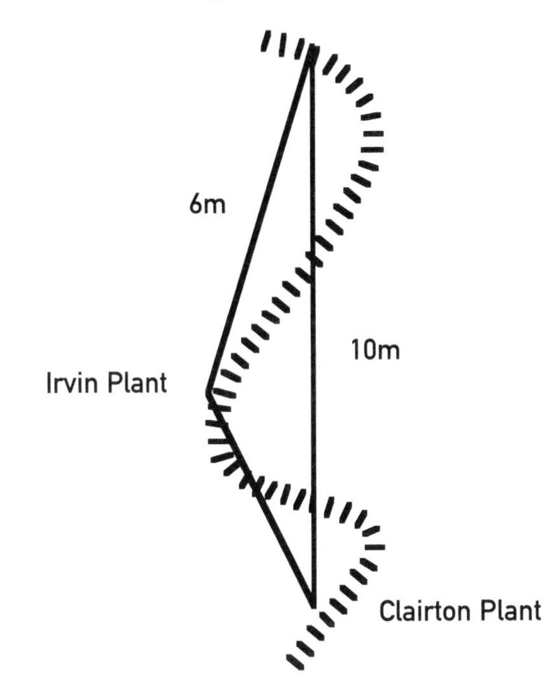

Figure 3.2. The location of U.S. Steel plants in the Pittsburgh area.

efficiency of each spill over to define the technology and efficiency of the whole.

The structure of management in these firms reflects the complexity of the production process. Because the individual plants are large, each has a separate internal organization and management hierarchy, and these in turn report to a corporate hierarchy. Historically, for the integrated companies decision making has been highly centralized. Information flowed ever upward and decisions, large or small, were pushed in the same direction.

Scrap-based producers typically began as single plant operations, although as they prospered they also grew and the number of plants in some firms has increased dramatically. The steelmaking operations in these firms do not rely on other plants owned by the same firm to supply the scrap metal charge; rather, it is obtained from independent scrap suppliers. Because of this there is obvious simplification on the input side of production, and the transformation of scrap metal to new steel is one process, uninterrupted between melting, casting, and rolling.

Management in these plants is lean and nonhierarchical. Communication is open and decisions are made in a context where all affected parties are in close working relationships. As these firms have grown, their management structure has remained simple. Decentralization of decision making has been their hallmark, and perhaps because their culture was shaped in single plant operations, they have eschewed bureaucracy as they have grown.

Summing Up

Picking the right market at the right time by seizing the opportunities presented as technology advances may sound simple, but it is not. There is much more to it than meets the eye. In this or any other industry, it is entirely possible to do everything right given the information at hand and still find yourself on the losing side. Getting it right, however, means not only making sound strategic moves in terms of market entry, but focusing the resources at your disposal to secure the market. The resources we are talking about are capital and labor, of course. Investments have to be well coordinated and their effectiveness has to maximized. People must be motivated, and they must be committed to goals that are articulated clearly. Moreover, people cannot be hamstrung by rules that limit their ability to contribute to the fullest. Success requires leadership and cooperation.

These elements in the struggle for market position among steelmakers are well exposed by the successes and failures that lay behind the industry as we know it today. In the next several chapters, we will draw upon the remarkable experiences of several firms to demonstrate the basis for success in the marketplace. In the process, we shall learn how the organization of the workplace, as defined by the technology employed in production, provides the framework by which management must translate its goals into reality.

4

Market Focus and Strategic Success

In the competitive struggle that we have described, some firms stand out from the rest as examples of achievement. Among the traditional integrated producers, the industry leader, United States Steel Group (USS), a division of USX Corporation, exemplifies successful restructuring. It eliminated inefficient mills or whole plants, got rid of unprofitable businesses, cut out layers of unneeded management, invested strategically in new technologies, and found the resources to carry out these needed investments. By the early to mid–1990s, this firm and a few other industry giants had been recast. Their gains were achieved at a tremendous cost to both workers and communities. But while the decisions involved were difficult and costly, they were necessary to the survival of the companies. Most were operating at less than 50 percent of capacity in the early 1980s, posting huge losses, and were not equipped to compete against imports or expanding minimills on either a cost or quality basis for many product lines. USS made the transition from this sorry state to an efficient producer of quality steel, and it made the transition especially well.[1]

As we have seen, however, it would be mistaken to view the transformation of the steel industry only in terms of the revitalization of Big Steel; the strategic gains posted by firms in the minimill sector have been central to the industry's newfound success. Those smaller companies, particularly the higher performing ones, provide valuable insights into what it takes to manage well in today's highly competitive marketplace. The ways in which the best companies in this sector achieved their successes show that management, through its values and practices, can create the culture and the environment in which all parts of the organization pull together to reach common objectives.

Among the minimills, Nucor Corporation is the clear leader. Nucor is the largest electric furnace-based steel company in the United States with sales approaching $3 billion and annual production exceeding 7 million tons. The company is the most entrepreneurial and innovative steelmaker in the world, with a consistent record of investing in new technologies to lower operating costs and move into more sophisticated, higher value-added products. Its financial performance consistently leads the industry. It has been profitable since the mid–1960s and in the last five years it has doubled in sales and almost quadrupled its profits.[2]

This chapter draws on the individual experience of USS and Nucor to examine carefully the first, and most important, element in achieving competitiveness: the strategic decision of market focus. In the case of USS this involved dramatically narrowing the scope of products in the company. For Nucor, it involved a number of sequential steps to expand market share and secure its gains. In subsequent chapters we shall return to these companies and others to capture the richness of how all the pieces fit together, as we examine other crucial elements in the competitiveness equation.

Common Ground

Successful minimills, like Nucor, are market focused, and the needs of these markets drive their investment decisions. When they enter a market they do so based on a decided cost advantage. In general their plants are small with few management layers; decision making is quick, with the participation of all who will be affected by the decision. Employees are expected to perform, and they are held accountable for performance. Communication is stressed and most information is readily available. Teamwork is the norm. These companies are focused, motivation is high, and the philosophy of continuous improvement is embedded in all aspects of their operations, including the ways in which employees carry out their jobs.

United States Steel contrasts markedly with the minimills in many ways. It was the dominant steel company in the United States for decades. When it was formed in the early 1900s, and for many years thereafter, it accounted for more than half the steel produced nationwide. This gradually declined, and by the 1980s the company's share of U.S. production had fallen to 20 percent. By the early 1990s its share was only 11 percent.[3]

Recasting USS to stem the losses and make it into a competitive steel company was not an easy task. Its culture, formed through decades of dominance in the U.S. marketplace, emphasized the status quo. Production men ran the company in the critical period of the mid–1960s through late 1970s and they regarded tonnage as the key measure of corporate performance, not quality or customer service. The organization was inbred, centralized, and autocratic, an environment not conducive to innovation or individual initiative. In the late 1970s the company did not have a clear focus on markets or products; it had a supermarket mentality, offering all types of steel for all purposes. The environment was such that production levels were reviewed daily on the sixty-first floor of the corporate executive offices in Pittsburgh and routine operating decisions were often made by corporate executives. There was little incentive for individual initiative or risk taking.

Given this history, however, there is surprising commonality in the basis for success at USS and the high performing minimills. Whether one turns to strategic decisions on market focus, the appropriate application of technology for the markets served, the critical role of leadership, or the overarching importance of productivity in capital and labor, the parallels to be drawn between Big Steel and its challengers are important. In these similarities lie critical elements of competitive success. Yet their significance is often lost on observers of the industry's performance in their rush to lay blame for failure and to claim knowledge of the keys to winning.

Defining Markets

Markets can be defined in terms of geography as well as products. For the steel industry it is impossible to grasp the essential character of restructuring without considering industry dynamics in both these dimensions, for they are closely related. The spatial factors involved in industry adjustment are well exposed by examining the geographic distribution of steelmaking capacity in industry subsectors. Let us begin by looking at a map of U.S. steelmaking capacity in 1974. Figure 4.1 displays the size and type of steelmaking capacity throughout the United States in that year. It shows clearly the dominance of integrated producers. Their furnace capacity in 1974 was 158 million tons and the product lines of these firms were broad.[4] They produced for every conceivable segment of the market, from construction and machinery to appliances and automobiles. Moreover, their competitors among the minimills were few in number and widely scattered. The minimills accounted for 8 million tons of furnace capacity, and this was highly concentrated in long products—primarily going to the construction industry.

The locations chosen by minimills in the mid–1970s reveal one aspect of their competitive advantage at that time. Of course there were exceptions that could be explained by special circumstances, but minimills most often sought advantage by serving markets that were at the very periphery of the market areas that could be challenged by integrated firms. The reinforcement bars, small beams and angles, and roofing supports that these firms produced had very low margins for profit. That meant that the shipping costs associated with getting these products to places like North

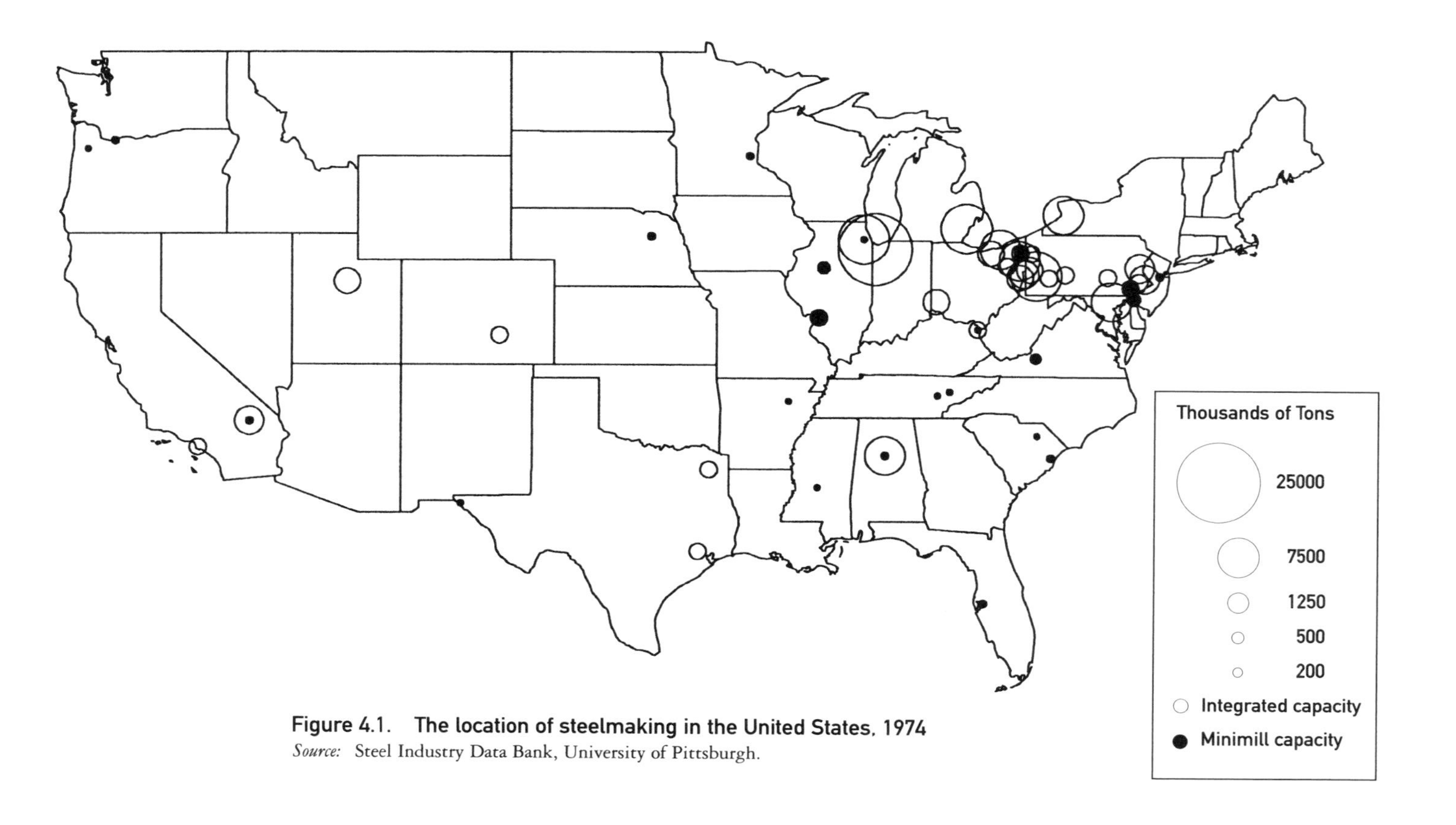

Figure 4.1. **The location of steelmaking in the United States, 1974**
Source: Steel Industry Data Bank, University of Pittsburgh.

Carolina, Georgia, or Tennessee from Big Steel concentrations in Pittsburgh or Chicago would put a minimill producer in the South at a real advantage, if their production costs were also low. Minimills had just that advantage, and they exploited it fully. These firms were located in parts of the country where population growth was rapid and the construction industry was booming, so the market for their products was strong. Their competitiveness was enhanced by substantial savings in capital costs, low electricity rates, and favorable access to their key material input, scrap metal. They sought and took advantage of locations that were outside the normal scrap supply areas used by the integrated producers. Because scrap metal was in surplus in these locations, this tended to dampen local prices. The other critical locational determinant for these firms was the local labor supply. They chose nonunion areas—specifically right-to-work states. Once there, they got highly motivated, highly capable trainees, who were most often new to the steel industry, and the companies used these workers well.

By 1994, the map of steelmaking in the United States had been transformed, as is depicted in Figure 4.2. The first and most obvious change that can be seen in a comparison of the two maps is the expansion of minimill capacity and the downsizing of capacity in integrated mills. But there is a somewhat more subtle difference. By 1994, the minimills were challenging Big Steel on its own turf; they were no longer seeking "protection" by serving local markets with a geographic advantage.[5]

Having proved that they could achieve a significant cost advantage over integrated producers, the scrap-based minimills challenged Big Steel on a very broad front. The market in low value-added products neared saturation, and profit margins there became very thin. As a result, the minimills began expanding product lines, which meant moving to higher value-added types of steel. However, this change reduced the value that firms attached to locations that were protected by distance from competitors. Steel products with higher value-added normally serve national markets, not local ones. Thus integrated mills were used to shipping these goods over great distances, and the minimill advantage in such product lines became even more dependent on their ability to stake a claim as the "low-cost" producer.

There is a fascinating parallel to be drawn here to the argument that is often made in international trade for small countries to protect an "infant industry" while it grows. The nurturing involved comes in the form of protective tariffs, but once the industry becomes mature and has gained the ability to produce efficiently by world standards, the tariffs can be reduced or abandoned because protection is no longer necessary. In the United States, the infant minimills became capable of competing globally. Minimills had

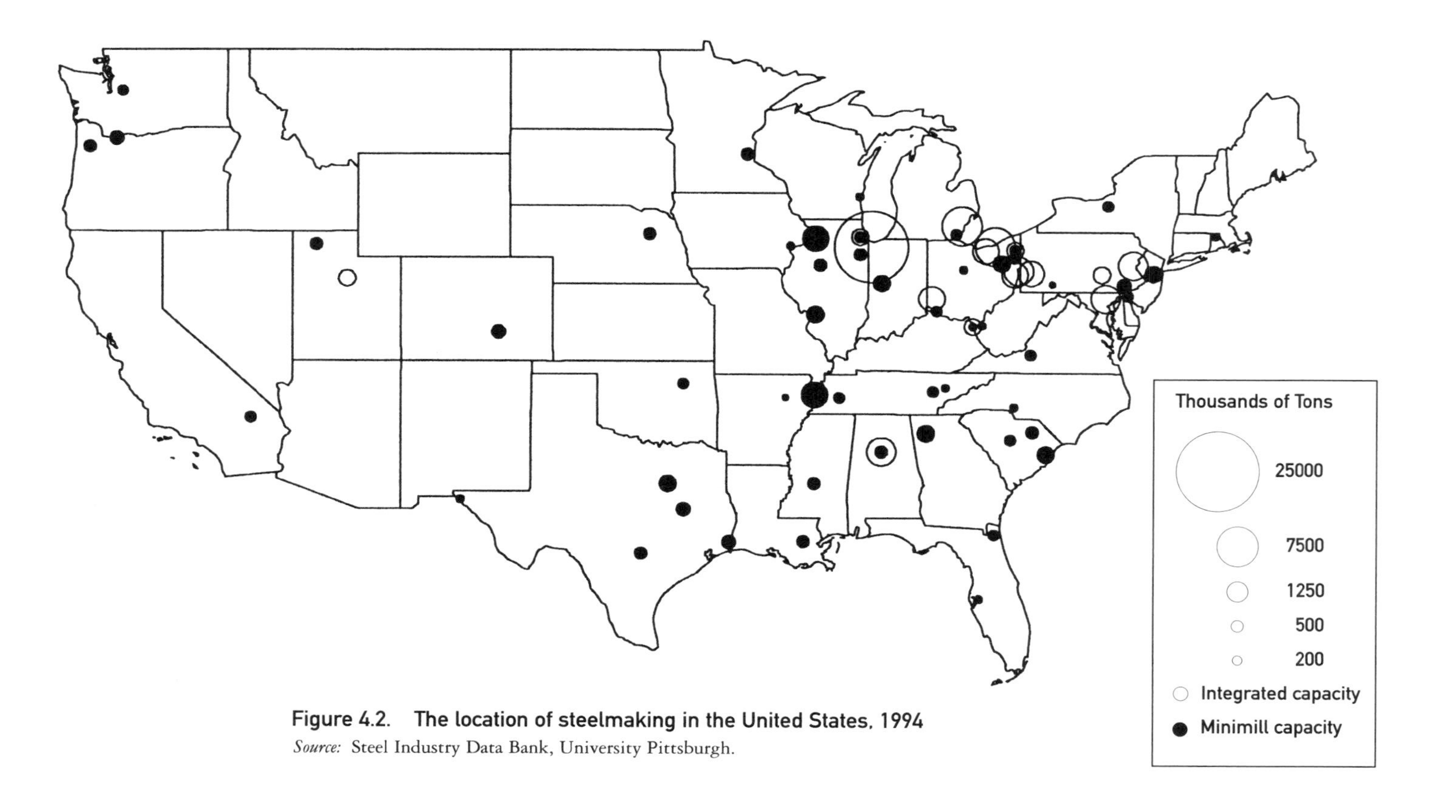

Figure 4.2. The location of steelmaking in the United States, 1994

Source: Steel Industry Data Bank, University Pittsburgh.

no government protector, but location and production cost advantage combined to offer barriers to competition from Big Steel. Both with respect to technology and management practices, knowledge gained in production at that early stage helped these firms prosper, so much so that one of their early protective barriers, transportation costs, substantially diminished in importance.

While the minimills thrived, Big Steel reeled under the strength of its competitors, domestic and foreign. Strategic decisions were required that involved identifying steel markets where ore-based production still had an advantage. Company by company, Big Steel firms came to the same conclusion: Markets for flat-rolled steel were the core of their business. When restructuring was initiated in the early to mid–1980s, scrap-based minimills were unable to penetrate this market and the barrier to entry was technical. The continuous casters in operation produced a slab eight to twelve inches thick, and they were expensive; moreover, the associated rolling operations were substantial. Outlays were required for a "primary" rolling mill that brought the slabs down to three inches. From there, secondary mills were required to finish the job. Between the capital costs associated with the caster and the rolling mills, slab casting required substantial scale for efficient operation in order to spread unit costs. Minimills simply did not have that scale.

The plants that Big Steel shut down and the capacity that it eliminated in those plants that remained primarily served markets for long products, where the competitive threat was most direct and profit rates were extremely low or negative. But plants serving flat product markets were closed or downsized too. They were characterized by high production costs that often stemmed from old or poorly utilized technology. Moreover, geography played a role in the decisions. Plants in close proximity to minimills and those closest to ocean ports (and, therefore, most vulnerable to imports) were at a decided competitive disadvantage, and suffered disproportionately because of this. Big Steel became highly concentrated in the market for flat products and in locations in very close proximity to the Great Lakes.[6]

We seek insight on these developments by closely examining the decisions taken by key players. The market growth of minimills and the narrowing of focus by Big Steel expose successful market strategies very well.

Strategic Market Focus: Nucor Corporation

The genesis of the Nucor Corporation was a company that was involved in the nuclear instrument and electronics industries in the late 1950s and early 1960s. Ken Iverson was installed as president in 1965 when the company was facing bankruptcy. The change in management led to a restructuring, and the decision was made to build the company around the profitable operations, which were those making steel joists. Even in the early 1980s, the company was still concentrated in two basic businesses, steelmaking and the production of steel joists. Today, Nucor has about 5,900 employees operating facilities in twenty locations. In general none of the plants employ more than five hundred workers. Together, they serve a very broad range of markets.[7]

Four Nucor mills produce bars, angles, and light structural carbon and alloy steel; two produce sheet steel; and one produces wide-flange steel beams and heavy structural steel products. The other divisions produce steel joists and decking, screen, bearings, and wire. About 75 percent of the steel mills' production is sold to outside customers. The balance is used internally by its other divisions.

In recent years, Nucor has diversified its product line to include a broader range of chemistries and sizes of coiled sheet, angles, rounds, channel, flats, forging billets, and special small shapes. These steel products have wide usage, including pipe, farm equipment, oil and gas equipment, mobile homes, transmission towers, bed frames, hand tools, automotive parts, highway signs, building construction, machinery, and industrial equipment. Customers are primarily steel service centers and manufacturers. Nucor's pricing structure is simple. All customers are charged the same published price, and prices are set to enable the plants to run at full capacity.

Strategic Investment

Nucor's performance has been driven by a dynamic investment program, of which its joint venture with Yamato Kogyo, a Japanese steel company, is characteristic. Together, in 1988 they completed construction of a new steel mill to produce wide-flange beams, pilings, and heavy structural steel products near Blytheville, Arkansas. The joint venture has a one million ton capacity and uses a continuous casting technology developed by Yamato Kogyo to produce a beam blank closer in shape to that of the finished beam than traditional technology. The operation of the plant is left exclusively to Nucor. This innovative technology, which is referred to generically as "near net shape casting," enabled Nucor to move successfully into a new market, previously the domain of the integrated mills, and to achieve the lowest costs among domestic producers.

In explaining how they got into this venture, Nucor's President, John Correnti, said,

> Yamato, through the Industrial Bank of Japan, approached us to help them manufacture in the U.S. because they saw their share of the U.S. market declining as a result of trade restrictions. I took a couple of trips to Japan to look at their mill. We came back and did our own calculations and concluded that the returns looked pretty good. The competitors were weak. So we jumped into it with both feet and off it went. It's been very very successful for us. The Japanese have 49 percent of it. They have the technical assistance agreement and Nucor has the management, the operation, and the construction agreements.[8]

The Yamato joint venture typifies Nucor's investment strategy of using new technology as a vehicle for market entry. But Nucor's most dramatic and visible success of this sort came in 1989 when it pioneered the implementation of thin-slab casting technology by constructing a one million ton plant at Crawfordsville, Indiana. This plant was followed in 1992 by another one million ton thin-slab facility at Hickman, Arkansas. Both operations (Crawfordsville and Hickman) are now being expanded to two million tons, and a third thin-slab plant is planned for Berkeley, South Carolina. These mills produce hot-rolled and cold-rolled steel and give Nucor the ability to compete for markets in the lower end of the quality spectrum for flat-rolled sheet with a cost structure significantly lower than the integrated companies.

Nucor's decision to invest $275 million in the new technology and the plant at Crawfordsville was reflective of its approach to risk taking. Early in his tenure as president, Ken Iverson "bet the company" when he borrowed $6 million to build Nucor's first steel mill in 1968.[9] Default on the bank loan would have bankrupted the company, but Iverson took the risk in order to lower the raw material costs for his joist business and to secure a stable source of steel supply. He also wanted to get into the steel business, and believed he could produce steel more cheaply than the integrated producers at the lower end of the market. He equipped the plant with the best technology available. While a major undertaking, the Crawfordsville decision was nevertheless just one step in the evolution of the company's philosophy of driving down costs and upgrading its product mix to stay ahead of its competitors.

Penetrating the Thin-Slab Market

In the mid–1980s, Nucor's markets were coming under increased competition from other minimills and management was therefore casting about for ways to change its product mix in order to compete against the integrated mills in markets out of reach of its smaller minimill competitors.

What it needed was a technology that would enable it to cast a thin slab that could be rolled into sheet steel without requiring the billion dollar investment in rolling mill facilities that traditional technologies required.

In the spring of 1986, Nucor executives visited a German foundry owned by SMS Schloemann-Siemag to inspect the company's compact strip production process, capable of casting steel in two-inch thicknesses instead of the industry norm of eight to twelve inches. If such a process could work commercially, it would provide significant production savings. The big question was whether it could be scaled up to provide the volume of production necessary for these savings and whether it could provide the quality necessary for the marketplace. None of the other companies inspecting the process thought the risks were worth the investment, but Nucor saw this as an opportunity to leapfrog the competition. They believed they could make it work, and they did. In December 1986, Nucor entered into an agreement with the German company to purchase what would become the first commercial scale thin-slab casting process.[10]

Soon after Nucor's decision, USS made a large investment in a traditional slab caster for the Edgar Thomson Works in Braddock, Pennsylvania. Like other integrated producers worldwide, they chose the conventional technology because the SMS process could not achieve the scale needed in their existing plants (e.g., three of the new, untried, SMS casters of the size ultimately used by Nucor would have been necessary to serve the blast furnace at the Edgar Thomson Works), and they saw real limitations to the quality of steel that would be produced from the new process. As important, there were logistical problems involved. To make maximum use of the SMS process, the thin slab it produced would have to be rolled while the strip was still hot in a continuous process when it left the caster. Final rolling facilities were already fixed for the steel being produced at Braddock, six miles away at the Irvin plant near Dravosburg, Pennsylvania. For other integrated producers the logistical problems may not have been this extreme, but every one of these producers saw limits they could not overcome or risks they could not bear. Nucor did not see limits; it saw potential.

The first heat of steel was successfully poured in the Crawfordsville plant in 1989, and the process has evolved greatly since then. While Nucor started by producing flat-rolled steel of the lowest quality type, they soon began to push their own envelope on quality outward, and penetrated the flat product market deeply as a result. This drove the price downward in some of the product markets served by the Edgar Thomson Works and similar integrated operations. The big initial advantage of thin-slab casting at Crawfordsville stemmed from lower capital costs because the need for a primary rolling mill is eliminated by casting thin in the first place. As important as this advantage is, secondary benefits followed immediately at Nucor. Improvements in operating and melting practices translated into gains in productivity and quality that opened the door to new customers and greater market penetration. New strategic investments came next in a cold-roll mill and a galvanizing line, further expanding the product mix. Moreover, in 1994, the company announced that it was expanding into stainless steel with the installation of a new refining furnace that would be used to reduce the carbon content of steel and add needed alloys. This is yet another step to move into higher value-added products.

When Nucor opened the second thin-slab plant at Hickman it was not content to simply duplicate its success at Crawfordsville. Rather, the company wanted to surpass it by pushing technology to the limit in order to open new markets. They adopted a new furnace technology, DC furnaces, at Hickman. This provided an important cost advantage for them. Nucor also pushed the casting and rolling technology at Hickman, and the result was that wider and thinner bands were achieved along with marketing and cost advantages. For example, the wider the steel is cast, the larger the diameter of the pipe that can be fashioned from the metal band. Nucor's experience at Crawfordsville made it possible to seek modifications in the funnel mold of the SMS caster that broadened the market that they could serve from the Hickman plant. Similarly, modifications incorporated into a newly planned Nucor mill in South Carolina will be based on the Hickman experience.[11]

Nucor's recent investment in iron carbide is another good example of its use of a new technology to gain strategic advantage. According to Correnti,

> We read about iron carbide in trade journals and it looked promising so our manager of technology went to Denver and talked to the inventor. He was impressed with the process and brought back data from an Australian pilot plant. We needed this low residual scrap substitute for our flat roll mills at Hickman and at Crawfordsville, and so we did some preliminary engineering on the process and determined we could do it for $60 million. We decided to go ahead and place it in Trinidad because that country has low gas costs. We're taking a $60 million risk. People ask me all the time, "What if it doesn't work?" I say, "We're not betting the company on it, but you know it's like anything else, little risk, little reward; big risk, big reward."[12]

In the event, the costs for this iron carbide facility may well run to $100 million or more, but in relation to the value of the potential market in flat products that is opened to them by this investment, it was still a very good gamble.[13]

A Forward-Looking Strategy

Nucor proved that thin-slab casting could work and that it was very profitable. As a result, tremendous interest has been sparked in thin-slab casting processes, and a number of new thin-slab ventures have recently been announced. Gallatin Steel, a company jointly owned by Dofasco and Co-Steel (both are Canadian firms), opened a one million ton facility in the United States in 1995 using virtually the same technology as Nucor. That facility can be expanded to 2 million tons. Steel Dynamics, a new company formed by former Nucor executives, will open a plant in 1996. North Star and Broken Hill Proprietary (Australia) announced plans to build a 1.5 million ton plant in the Midwest to open in 1996. LTV, British Steel, and Sumitomo Metal Industries recently announced that a major 2.2 million ton plant is to be built in Tennessee or Kentucky (Trico Steel). World Class Steel is raising the financing to build a new plant. Nucor has started construction of a third flat-rolled plant in South Carolina. Acme Steel is installing a thin-slab caster that will be using metal from a BOF. By the year 2000 there will be at least 12 million tons and possibly 18 million tons of new flat-rolled capacity since the "Nucor Revolution."[14] This will put substantial downward pressure on prices in the product markets involved and substantial upward pressure on the price for scrap metal—especially high-quality scrap.

Nucor's initiative in iron carbide looks ahead in order to protect the market position it gained with the Crawfordsville and Hickman operations and to extend its advantage. Iron carbide from Trinidad will be used as a substitute for scrap metal in Nucor's flat-rolled operations. The advantage is twofold: First, using iron in the furnace charge at these plants diversifies the product mix and makes Nucor less vulnerable to the threat of higher scrap prices. In late 1993 and in much of 1994 the price of higher-quality scrap was close to $150 per ton, about a 50 percent increase from early 1992.[15] At this price the production cost of flat-rolled steel from the EAF thin-slab caster is less attractive and approaching that of the integrated producer. By increasing the amount of iron used in the charge, Nucor can substitute lower-quality (less expensive) scrap for high-quality (more expensive) scrap. So if the technology succeeds at Nucor's iron carbide plant, it secures the firm's position as the low-cost producer in this market. Second, by mixing greater amounts of iron with scrap inputs, Nucor can penetrate the flat-product market more deeply by going after market segments that demand somewhat higher surface quality—markets that will be denied to electric furnace producers that rely more heavily on scrap metal inputs.

Strategic Market Focus: United States Steel

During the 1980s, the integrated companies tried a number of strategies to improve their performance.[16] There were wage and benefit givebacks by union and hourly workers that amounted to $2 billion in 1983 alone, and this was followed by more givebacks in 1986.[17] Work rules were loosened and job classifications made less restrictive to give the companies greater flexibility on the plant floor, thereby increasing efficiency. Big Steel sought joint ventures and alliances with foreign steel producers. Capital investments were made to increase the efficiency of existing facilities. Firms diversified to other industries in attempts to improve financial performance, and they reorganized and focused steel operations.

While some of the above strategies brought temporary relief to the industry—the wage concessions and increased productivity lowered costs and the joint ventures and alliances brought in investment capital—none mitigated the need to scale down operations and to better manage what remained. United States Steel was at the eye of this storm.

David M. Roderick was appointed Chief Executive Officer (CEO) of USS in 1979, taking over from Edward Speer, an autocratic executive who had a strong production background. Immediately, Roderick set about closing some of the company's unprofitable plants and inefficient mills in its major facilities and formulating a diversification strategy. By late 1979, fifteen facilities were closed and eleven thousand workers laid off,[18] but more drastic action was called for in the early 1980s. Marathon Oil was acquired in 1982, which greatly decreased the corporation's reliance on the steel business. But the magnitude of the steel problem escalated (the name of the corporation was changed to USX that year). In 1982 losses in the steel business accelerated and production plummeted, producing a capacity utilization rate of 36 percent. An operating loss of $850 million was recorded that year, to be followed by a loss exceeding $600 million in 1983. During this period the company was losing up to $75 on every ton of steel it shipped.[19]

As part of a corporatewide strategy to deal with the steel division's problems, in May 1983 Roderick recruited Thomas C. Graham (Tom Graham) to USX as Vice Chairman of Steel and Related Resources and President and CEO of the U.S. Steel Group. He had been President and CEO of Jones and Laughlin Steel Corporation at the time of his recruitment, and had the reputation of being the most competent steel operating executive in the industry.

During his eight-year reign as CEO of the U.S. Steel Group, Graham turned the company into the top performing company in the integrated sector. He did it through a management style that was both combative and irreverent. He had a vision, communicated it throughout the organization, and saw to it that it was implemented. He had little patience for bureaucracy and saw himself as an agent of change. Called the "smiling barracuda" for his ability to make tough decisions, Graham's management style produced exemplary results at all the companies he led.[20]

A Sense of Urgency

Shortly after arriving at USS, Tom Graham was given the profit and loss statement (P&L) and was stunned to see that the steel division had lost $100 million during the month of May. Graham stated, "With losses like that, you can't meditate for a year on what you have to do. There's a certain urgency, because we were 'bleeding cash by the bucketful.' You couldn't shut the money-losing operations down, because everything was losing money!"[21]

USS was losing money not only on a full-cost basis but on a variable-cost basis as well. Graham said: "In truth, the P&L isn't much of a guide as far as providing insight into what should survive and what should go. You have to make a judgment at a time like that on what operations should survive, which ones have the long-term potential to be profitable, and which offer little or no potential for the future. No computer program has been devised to solve these problems. It's purely a judgment call."[22]

Defining the scope of the problem, Graham noted that USS required about eleven man-hours per shipped ton, greatly exceeding the productivity levels that he had achieved at J&L. At that time, USS was also producing a number of product lines that competed with scrap-based (minimill) producers. There was almost no way the company could ever be price competitive in many of those low value-added product areas. USS had too many plants, too much equipment, and had spent capital everywhere, but there was no long-term strategy guiding its decisions. For instance, at the Cleveland Cuyahoga Works, which was in the rod business, the company had invested significant amounts to upgrade the facilities and was never able to compete on a cost basis with the minimills.[23]

A Market-Driven Strategy

When the restructuring of USS was completed six years later, the company was basically a flat-rolled producer serving primarily the automotive and appliance industries. In describing how he formulated his strategy, Graham stressed, "It was market, market, market driven. We had to respond to our customers in our desired markets with quality products at a competitive price. That factor influenced our decisions on investment strategy, equipment, facilities, and the people who managed them."[24]

Therefore, once the decision was made to serve certain markets with specific products, the decisions to close some facilities and invest in others were readily apparent. And since Graham came from outside the corporation, he was not constrained by its history. He looked objectively at what needed to be accomplished in order to make money in the markets served. He developed his own vision for the company's future and the strategy for carrying it out.

Some of the decisions were easier than others. For example, in 1983, USS was still making open-hearth steel at the Homestead mill outside Pittsburgh, and the finishing end at Homestead had two obsolete plate mills and two 1918 structural mills, all supplied with high-cost steel from the open-hearth shop. This plant also had a cavernous forge shop that was the only domestic producer of propeller shafts for aircraft carriers, but Congress had not authorized any funds for new carriers in some time. The decision to close Homestead was not a hard one to make. Nor was the decision to close the Duquesne Works, located next to Homestead, complicated to reach. Duquesne was competing in the bar business against the minimills. The newest bar mill in the plant had been built in the 1920s and there was no way to get a decent product off that equipment and compete with mills that made a better product at a lower cost. Duquesne did have a relatively new blast furnace, but it did not matter how good the furnace was, because the downstream units were so inefficient that the resulting quality and costs were not competitive.[25]

Similarly, the nearby McKeesport plant had obsolete seamless tube mills, and the company had a new mill under construction at another plant that was 95 percent complete. At that time, the seamless tube market was nearly nonexistent, so it was not a difficult decision to close an antiquated facility for one that was state-of-the-art. The company had an efficient welded pipe mill in McKeesport that was linked administratively to one in Lorain, Ohio, and ultimately that operation was spun off as a joint venture.[26]

Forming a long-range market strategy took time. Graham explained, "As my knowledge of the U.S. Steel organization grew, my vision for the company was to restructure it into a flat-roll company with the exception of the new Fairfield seamless mill and two bar mills at Lorain. We had a lot of other unprofitable facilities and we made a lot of products that weren't very good. They all had to go."[27]

Graham's decision was based upon the knowledge that the flat-rolled market offered his company the best return for the future as a consequence of the change in technology that had occurred in the 1980s. The large integrated multiproduct plants were a result of the traditional practice of pouring molten steel into ingots, which was the mainstay of the industry until the development of continuous casting. In a steelmaking shop that teemed ingots, the ingots went to either a slab mill if it was a flat-roll order, or to a blooming mill if the order was for a billet that could be made into bars and pipe. In that setup, the steelmaking operation was insulated from the finished product.

The world changed with the advent of continuous casting, which is a significantly more efficient steelmaking process. When steel is continuously cast, it is committed to a certain finished product. The marketing options are then more limited than with ingot teeming. Once committed to staying in the steel business by focusing on the market for flat products, the only way to remain competitive was to build a slab caster to achieve the quality and low costs necessary to meet the competition in the marketplace, and to achieve the high volumes required to justify the larger capital investments in the blast furnaces, coke ovens, and steelmaking operations.

Graham's approach to restructuring the company worked. By the time he left the presidency of USS, the company was profitable and its productivity had improved to the point where the labor hours required to produce a ton of steel approached four, down from eleven when he came on board.[28] The company primarily produced flat-rolled products. It was completely out of bars and rods, and soon to be out of the structural steel market as well. Moreover, about two-thirds of its steel was continuously cast compared to about 10 percent when he took over. During this period the number of plants was reduced from fifteen to five, capacity fell by about half, and employment was reduced by more than half.[29]

His strategy was very much based in common sense: know the markets you want to serve, reconfigure your assets to become the low-cost producer in those markets, and set about to improve the performance of the surviving mills. What sets him apart from other steel executives was his willingness to make the difficult decisions and his understanding of how to ensure implementation.

A Cost-Driven Strategy

Market selection initially drove the restructuring of USS, but once the decisions were made with respect to which facilities were to continue in operation, the emphasis turned toward making them as efficient as possible. Important aspects of Graham's decision making at USS are revealed by his later experience. Graham left USS in 1991 and soon after that joined Armco Steel which had just two plants, producing flat-rolled products. His primary need was to turn these into profitable operations. The company was already in the right markets, but the productivity of its existing mills was too low.

Graham explained his approach:

> The first thing I noticed when I reviewed the operations was that Armco operated too many units for its current and foreseeable business. If you aggregated all that business, we had enough to operate one fully loaded, good-performing unit. However, this was a two-plant company with a two-plant psychology and a two-plant mentality. Neither facility was ever going to achieve the economies of scale that are so vital if you are competing in the world steel market. For example, take a tandem cold rolling mill, which is an essential piece of equipment if you make flat-rolled products, like cold rolled sheet, galvanized sheet and so on. A world-class tandem cold mill makes 175,000 to 180,000 tons a month. Armco sold about 180,000 tons of cold rolled a month but the production was spread over two mills, one in Middletown and one in Ashland. The same situation existed with the hot strip mills. We had a great hot strip mill in Middletown, capable of world class production. The one at Ashland had outlived its usefulness.[30]

The Ashland hot strip mill and cold mill were eventually shut down, with the Middletown mills taking on the additional tonnage. This raised their output and productivity, thereby reducing costs per ton shipped.

After closing unprofitable plants and mills within plants, success depends upon being able to increase the productivity of what remains. This may require new investment in order to provide the tools to compete with the best in the world, but it also requires leadership that is committed to achieving higher levels of performance, and managers who are willing to change and do things differently.

As a general principle for restructuring a company, Tom Graham emphasized, "I always start at the market end. You want to see over the long term your company's place in that market. The market doesn't stand still. What you want to look for is a durable market, where you can reasonably expect to participate, if you are cost competitive and have good quality and service."[31] The importance that should be attached to this kind of focus is enormous. Capital expenditures need to be market driven in order for them to provide the greatest returns to the company and its customers.

In a large organization like USS, investment decisions can be bureaucratized easily, but when capital is scarce and the success of individual investment decisions can make or break a company waste cannot be tolerated. At USS there was an unfortunate history to the way investment dollars were allocated across plants, a history in which the balance in corporate politics was given significant weight. Graham changed this. As he explained,

> In most situations you have to be scrupulous about capital expenditures because cash is in short supply. You have to be very sure that you're spending money on a venture and a product that has a durable future, and that you are making a substantial contribution to that durability. The one thing a CEO can never delegate away is the responsibility for the capital, and I don't. You never have enough capital, and you have to be wary about spreading it around. As hard as it may be, you have to spend the money where it will do the most good and that is in the area of highest priority given the markets you serve. You can't throw it away out of compassion or a sense of equity. You can't run from the unpleasantness associated with turning people down, if it's the right thing to do.[32]

Like every other integrated producer, USS is now reassessing its position in the flat-rolled market in light of Nucor's success in thin-slab casting. For Big Steel, market advantage, at least for now, lies in the melt chemistry of steel produced from ore rather than scrap and in scale. USS, for example, has recently ordered a vacuum degasser for the Edgar Thomson Works that would position it in a market for ultralow carbon steels. Given present technology, steel produced from scrap metal with minor amounts of DRI or briquetted iron contains impurities as residual elements. The properties required for ultra-low carbon steels, mainly formability, cannot be achieved in scrap-based production even by reducing carbon in a vacuum degasser because the residual elements harden the steel. Moreover, the cost of vacuum degassers is substantial, and efficient operation requires large-scale production. So even if a minimill changed its input mix to use a substantial portion of DRI or another type of refined ore, it could not match the quality of steel being produced at Edgar Thomson and its scale would still be a limiting factor in competing in this market niche.

The Market Contest Today

With Big Steel having focused on the market for flat-rolled products and EAF minimills pushing ever forward into this market, the inevitable question is, "Who wins?" There are two certainties involved: (1) Many of the new EAF thin-slab producers that are following Nucor's lead will base their production on a furnace charge that uses scrap metal substitutes from virgin iron as well as scrap, and (2) as experience is gained in thin-slab casting, the quality of steel from this process will improve.

Right now, lower grade flat-rolled products can be produced primarily from scrap metal with as little as 20 percent DRI in the EAF charge and cast in a thin slab. Based on the information available today for prices and technology, and recognizing that process know-how will grow as experience in thin-slab casting accumulates, it is reasonable to believe that minimills could have access to as much as 15 to 20 million tons of a total flat-rolled market that is of the order of 40 to 45 million tons per year.[33] In the downsizing and restructuring of the 1980s a few integrated firms such as Wheeling-Pittsburgh Steel and McLouth focused specifically on that market, that is, lower grade flat products. Now they are going to have to find new market position or reduce their capacity further. Even the first order integrated firms such as USS and Bethlehem that have the capability of producing steel of the highest quality—quality that cannot be touched by the insurgent minimills—will be directly affected. In the normal course of production some of the steel they produce fails to reach necessary specification requirements, and when this happens it is downgraded to serve lower-end uses. Although downgrading of this sort can be minimized—and there is much less of it today than there was even a few years ago—it cannot be eliminated. As a result, depressed prices for lower grades will translate into depressed profits even for the premier integrated producers. Neither are imports exempt for the competition. The U.S. steel market is and will remain one of the most competitive in the world. Only the most efficient producers of any description will have access to it.

Right now and for the foreseeable future, the contest remains concentrated in the lower end flat-rolled market, where competition will be fierce for the next decade. The first 8 to 10 million tons of flat-rolled steel produced with thin-slab casters (Nucor's 4 million plus 4 to 6 million more) will penetrate that market fairly easily, due to the lower cost associated with the technology. Further penetration will occur but will be more difficult as increasing scrap prices reduce the minimill cost advantage and integrated producers keep the pressure on by realizing further efficiency gains. In this context, it is clear that firms participating in the flat-rolled market must have a well-defined strategy. Those simply attempting to copy Nucor may find initial success, but as the market gets more crowded profits will shrink or turn to losses.

Summing Up

We find in Nucor and United States Steel some practical rules. Focus on a market and make the investments that will allow you to capture that market by achieving cost and quality advantages that can be sustained. Simple enough, but doing it requires leadership that draws the best out of technology and people. Nucor's success is partly due to its technology, but leadership and culture really define its long-term advantage. And it is clear that the success achieved at USS required the ability of a leader to implement necessary reforms. Any producer in this market, minimill or otherwise, has to recognize this fact or it will not thrive.

In the next chapter we look at management issues much more closely, for the pieces of this puzzle are many and they fit together in subtle ways. Competitiveness requires attention to that detail.

5

Leadership and Management

Whether one examines the progress of USS or Nucor's enormous achievements, it is impossible to escape the fact that dynamic leadership is required to shape the critical decisions that focus a firm's resources and help it to gain competitive advantage in a market. Of course, even for the best leader, success requires a good deal of luck. It is entirely possible to make good decisions, given the available information, only to find that things go horribly wrong for reasons beyond the company's control. Once critical decisions about the firm's focus have been made, however, and an investment strategy is mapped out in order to serve that market, success often comes down to the effectiveness of management. This, too, cuts across Big Steel and the minimills. In this chapter, we look at the management side of the steel industry's turnaround, and once again the lessons generalize well beyond steel firms.

This chapter shows that strong leaders do not micromanage a company. Instead, they have a conviction of purpose for the company and its goals, and they communicate this broadly to gain the commitment of others. They set expectations for performance, provide people with the authority to meet agreed-upon objectives, hold them accountable for performance, and reward them accordingly.

These leaders are never satisfied with the status quo. They build a culture that is based on openness, trust, and a shared sense of the organization's purpose. This gives rise to decentralized decision making and encourages risk taking. As a result, innovation and acceptance of change are established as the norm, and employees are motivated to seek improvement in performance both in terms of their own job and of the firm itself.

Goals Must be Held in Common

It does little good for a CEO to know where a company has to go if he or she cannot get the people that comprise the firm to share in that commitment. They must *want* to go there in the sense that they perceive that their well-being and the company's are one and are convinced that both will be better off if certain goals can be achieved. Overriding all this, however, is the need for leadership that brings coherence to the enterprise. The CEOs in major steel corporations speak of the importance of having a "vision" of what the company is and what it is going to be. That sense of purpose is the foundation of successful management.

Some decisions, the big ones, are taken alone, and ultimate responsibility rests at the top. When Tom Graham looked at the facts facing USS and decided that resources had to be focused on the market for flat-rolled products, plants had to be closed, and there was no alternative. Later, when he was CEO at Armco Steel (now called AK Steel Corporation), he knew that major facilities at one of the company's two plants had to go. At neither of these junctures, could responsibility for his strategic choice be shared. If he succeeded, the company succeeded; if he failed, the company failed. These were not decisions that required consensus in any meaningful way. Similarly, when Ken Iverson, Chairman and CEO of Nucor Corporation, recognized the opportunities that existed if his company moved beyond steel joists and rebar, he may have sought advice broadly but the decision to seize new markets was his to make.

Implementation of those decisions, however, depended on other people doing their jobs well, and failure to recognize that fact could easily have cost Graham or Iverson and their firms dearly. For Graham at Armco, the decision to close the hot strip mill and the cold strip mill at the company's Ashland Plant was also a decision to make the company's Middletown plant into a world-class operation. The decision to close facilities or shutter a plant cannot be an "easy" one; personal hardship will ensue for hundreds or even thousands of people once it is taken. But the closure decision did not require cooperation; it required conviction. In contrast, bringing Middletown to its maximum potential in terms of productivity and quality required cooperation.

Speaking of the task at Middletown, Graham put it this way:

> The Middletown hot strip mill had been running at 280,000 tons a month for years, but I knew it was capable of operating at levels well in excess of 380,000 tons a month. The first step was to sell management on that objective, realizing that it could not be accomplished quickly. It took over a year and was a result of hundreds of different improvements: purchasing harder rolls to keep the mill working longer between roll changes, improving maintenance practices to reduce the number of breakdowns, improving roll changing practice to reduce the time between roll changes and so on. We had to create an environment in which all levels of management were committed to achieving this objective. One superintendent had to be

> reassigned because he was not able to accomplish what we thought was achievable, and of course this helps to communicate the seriousness of the task.[1]

The commitment of all levels of management was important because the changes that were required ranged throughout the plant. Opportunities for improvement in plant operations, which might exist in any number of places, had to be recognized. No CEO could know them all, nor would it be reasonable to expect one to do so. This means that information has to be obtained on where the opportunities lie; priorities have to be assigned in making those improvements, especially if new financial resources are required to implement them; and someone has to be responsible for each improvement—responsible for its success or failure. None of this can be accomplished without decentralization, nor can it be accomplished well unless people are well motivated.

Good managers know this, and they know that the need for decentralization is not limited to the ranks of management. John Correnti, President of Nucor, put it this way:

> I can't melt steel or roll steel or sell steel or account for steel as well as those guys in that plant who do that for a living. A lot of people think that because they have the title president or executive vice president, they know more about the business than the guys on the shop floor and that's not true. I know more about the general part of it than they do, but the melter knows more about melting steel than I do, the roller knows more about rolling steel, and the sales people know more about selling steel. So you give them the encouragement to do their jobs to the best of their ability and you push it downward.[2]

Top management needs cooperation from every level of the enterprise if a firm is going to realize and sustain a competitive advantage. This attitude is central to success in integrated firms and in the minimills, and is widely practiced in the best firms. But the way it is implemented and the limits involved are highly dependent on the structure of the firm. By drawing on actual practice, we hope to expose the key elements of success in managing steel firms—elements that are widely found in other industries too.

Keys to Successful Leadership: Drawing on Graham's Experience

In any integrated steel firm, there is an enormous separation of workers and management that stems from the organization of production and from a long history of adversarial labor relations in the industry. As we have explained, integrated production is large in scale and, typically, organized in multiplant firms. In these plants, a steelworker's first loyalty is traditionally to the union, and herein lies the gulf between management and labor. Because of this, one moves an integrated steel firm forward by engaging management first. Those who are going to carry out the program for change envisioned by the CEO must be committed to change themselves: They must "buy in," or the opportunity for success will be lost.

We can use Tom Graham's experience at Armco to drive this point home. Recall that Armco's Ashland plant had antiquated, inefficient hot strip and cold mills and the Middletown plant was going to pick up the entire load, which meant increasing production at Middletown enormously. With Ashland in full operation, Middletown had been making 225 tons per hour in its cold mill. Graham believed that if his restructuring plan was to succeed Middletown had to be brought up to at least 300 tons per hour, which would entail a 33 percent increase in productivity. The management at Middletown saw this for the challenge that it was, and no one was going to leap forward to say that the people and equipment they had in place could now be brought up to that level of performance. Tom Graham was convinced it could be done; indeed it had to be done, and the job would fall to managers at Middletown.[3]

Speaking about that challenge, Graham explained the steps involved this way

> It was the confidence of the management and crew that was holding it down. Instead of ordering 300 tons an hour, I went out to the mill and watched it run for a while. I talked with the superintendent about the mill's performance. I told him the performance was not what I would call a world-class operation, and I asked him to write me a letter detailing what he needed in order to achieve 300 tons per hour but without spending a gob of capital. He then met with the crew and his supervisors, and together they put together a plan to make it happen. What they needed to get started was some encouragement, and the knowledge that senior management wanted their help. For me to sit in my office and demand 300 tons per hour is a little naive. I have to get that commitment from the organization or it will never happen. The important thing to keep in mind is that I can't do this job alone. Other people have to pick up the ball and commit themselves to the game. They have to emotionally and intellectually agree that this guy Graham is after the right things. And I'm busting my tail to see that that happens.[4]

People cannot be asked to do something they think is impossible, however, and when a dramatic change is made in the status quo it may be difficult for many people to recognize a firm's real potential. Because of this, success in restructuring often comes down to helping others see and seize opportunities for improvement. Good managers recognize the constraints on performance, and work to relax them if possible. When the constraints cannot be relaxed, good managers are able to perform to the limit within them.

This process is perhaps easier to recognize in the steel industry than it is in many other manufacturing processes because of the sequential nature of steel production. From iron-ore pellets to the finished products being shipped from an integrated steel plant, one might identify fifteen or more steps such as ore processing, hot metal production, steelmaking, and rolling. Every one of these has a given capacity to move a certain amount of product in a day, which means that any one of these may be a constraint on the capacity of the entire system. In capital intensive production processes of this type, bottlenecks are potentially rich sources of productivity gain. Indeed, in the cash-starved steel industry of the 1980s when few firms could rely on major new equipment purchases to improve their cost competitiveness, eliminating bottlenecks was vital.

Bottlenecks can involve entire plants or small parts of a rolling or finishing operation. Top managers can identify and deal with critical relationships among plants, but they cannot reach down to the shop floor. Productivity gains based on relieving bottlenecks at that level must be recognized and eliminated by those most directly involved. This is one of the critical reasons why everyone in a firm has to be committed to success.

Tom Graham's approach to accomplishing this is both simple and direct. He recognizes the importance of leadership and of instilling a sense of direction and purpose in those who work for him. To accomplish this he gets involved personally by reaching out to those he believes are in the best position to know where opportunities lie. At USS he would meet with plant managers monthly to review performance and help identify bottlenecks. These meetings were taken as opportunities for exchange, not tests of loyalty or fealty. In them, he would seek opportunities for improvement and ratchet expectations for performance upward. But this had to be done in a way that retained and strengthened commitment. Benchmarking was a useful way of doing this. Very often the benchmarks were internal and set performance targets with respect to productivity, safety, or quality, but benchmarking against the competition was also used as a means of reinforcing the idea that the goals he established were, in fact, achievable. After all, if their competition could do it, why couldn't USS do it too? However, whether the targets were internal or external, the idea was the same: To get commitment and keep people focused on goals that contribute to financial performance.[5]

He carried the idea that commitment was essential with him to Armco Steel. In order to get the commitment he needed at the Middletown plant, Graham held weekly "prayer meetings" at a nearby hotel. Again, exchange of information was critical. He characterized the meetings this way:

> It's a cram course in getting to know how management thinks, letting them get to know me and having a free exchange of ideas. I have a different operating or staff group each time. Each person gives a 10 to 15 minute presentation about his job, and they unload all their bitches because they have me as an audience. I quiz them, and when the group is done, we have a freewheeling conversation about the company's policies and practices. We have dinner and go home. The virtue in this arrangement, since I reach down about three layers in the organization, is that they open up after some prodding and I learn things I might not otherwise hear. I am very judicious about how I use the information I get at these meetings. My principal motive is to sell my management philosophy and learn more about the operation from the people who are in the trenches.[6]

He approached every one of these meetings with a single objective: He wanted to get enthusiastic commitment from everyone.

In this process Graham was able to identify new leadership potential and capitalize on it. He arrived at Armco in January 1992, and by the end of 1993 seventy of the company's top managers had been replaced.[7] All but twelve of these were promoted from within. Graham was looking for commitment and drive, and he found it in people throughout the company. Then he translated their commitment into profits. In 1992, the company was losing about $50 for every ton of steel shipped. A year and a half later it was making $10 a ton while most of the other integrated producers were still posting losses, and by 1995 it was earning more than $70 for every ton shipped, the highest of the U.S. integrated producers. This was accomplished through a constant focus on productivity gains as seen in a reduction in man-hours per ton from six to less than four in mid–1994, and significant improvements in quality.[8]

Graham's leadership reshaped the companies he led. He was not constrained by past practices, but aggressively sought change. He implemented his vision by expecting those around him to meet difficult challenges. He kept the pressure on the organization through extensive communication, both formal and informal, and by quantifiable goals and objectives, which he monitored. Therefore people knew what was expected of them, and they knew he was serious. This drove the organization in the direction he desired.

Strong leadership is essential to changing the culture of a well-established company. In all the companies Graham managed, there were traditional ways of doing business which did not produce the results required to be competitive in an industry facing challenges from minimills, foreign competitors, and other domestic integrated producers. Drastic action was called for, and this required decisive management.

Keys to Successful Leadership: Drawing on the Minimill Experience

Minimills can be single-plant operations or, like Nucor, they may involve many plants. What they all have in common is the simplicity of the production process. Scrap metal goes directly to finished steel in a matter of hours (not days as in ore-based production), and the stages involved are limited. As the name suggests, minimill plants are also relatively small in size (typically one million tons or less). Many, but not all, began as "greenfield" operations (new plants on newly developed sites), which were spared the history of soured labor relations that has characterized Big Steel.

The size and character of minimill operations is important to understanding the effectiveness of their leaders in helping to transform the American steel industry. There is no doubt that technology gives minimills cost savings that they have used well in securing competitive advantage. But to listen to the CEOs at successful minimills is to appreciate that technology is only part of their success. As Nucor's John Correnti put it: "Technology accounts for about 30 percent of Nucor's success. The other 70 percent is our culture and employee relations/employee management practices."[9] The split may be 30–70 as Correnti believes, or 50–50, or even 70–30, but the point is that Correnti's sentiment is echoed widely by other industry leaders: People matter a lot.

In a typical minimill, getting the commitment of management is not a very difficult task. The operations of a given plant are small enough that all of the managers can (and often do) fit in one room. The implementation of a decision that requires coordination of effort across the plant might have as few as five or six people involved at the outset: the CEO, the chief financial officer, the person responsible for buying scrap metal, the manager of the melt shop, the manager of the casting and rolling operations, and the marketing person. That covers management! For coordination of decision making across an entire company in a multiplant operation, the mix of people involved would be different, but the number would not. In either case, a plant or a firm, the chances are high that all those people would have been involved to some extent in the initial strategic decisions about new investments, and that they already have a stake in the decision's success as a result. Communication with management does not pose the same kind of problem that Graham encountered.

Decentralize! Decentralize! Decentralize!

Success in the minimill depends on grassroots participation. Everyone has to be on board, management and workers, and they have to be driving toward the same goals with the same intensity of effort. That can only be accomplished if a culture that promotes high levels of motivation and cooperation throughout the organization is established and maintained. One of the keys to this is decentralization within the firm and plant.

Again, Nucor stands as an example of how this works. The firm is organized in seventeen production units or divisions, seven of which are steel mills. Each division runs as an entity unto itself, headed by a Vice President General Manager who is an officer of the company. Nucor provides capital, technology, know-how, and training, but critical responsibility for production, hiring, purchasing, engineering, safety, quality control, sales and marketing, credit and collections, and to some extent pricing reside at the division level. In turn, that General Manager puts the melt shop manager in a little business for himself, and the melt shop manager pushes decision making and responsibility down even further.[10]

What one finds at Nucor is that responsibility is widely distributed. For proof of this, one need only look at the staffing levels at its corporate headquarters. In Correnti's words:

> When we were a $1 billion company we had 18 people in our corporate headquarters. When we were $1.5 billion company we had 19. Now we have 22 people here at the corporate office. Having another division report to me means that instead of reading 17 lines on a piece of paper I have to read 18 lines. You have to keep the company decentralized, and you can't let yourself fall into the trap that says bigger means more corporate overhead. It has worked for us in the past and I'll fight at all costs to avoid building up a corporate hierarchy. It stifles growth, it stifles ingenuity, and it stifles that entrepreneurial spirit. There are times, however, when I'd like to reach through the telephone and grab that steering wheel and jerk it, but you can't do it. You have to influence, not direct or order, the guy through the telephone who has his hands on the steering wheel."[11]

In fact, the information provided to headquarters includes just a few sheets of paper: monthly operating reports, weekly tonnage reports, and monthly cash management reports from each plant. That is all the formal information they get about performance! The standardization of these reports and the simplicity of the company's low-cost, high-capacity utilization strategy enables corporate headquarters to push decentralization to the extent it has.

The limited amount of information provided and the decentralized decision making structure enables those at the corporate office to manage effectively with a wide span of control. But decentralization has another very important purpose. When people accept responsibility they do so knowing that trust has been expressed in their ability to carry through. Performance expectations may be set from above, but employees at every level of the firm are given significant discretion in carrying out their functions to accomplish these objectives.

Establish a Basis for Trust

There is an element here that stands out and seems to generalize for the most successful minimills, and this is that expressions of respect, trust, and confidence in one's employees go hand in hand with the ability of a firm to excel. To achieve this, many of the minimills have aggressively sought to break down the cultural barriers within firms that foster an atmosphere of "us versus them" between the ranks of managers and the workers on the shop floor. There are a number of examples that can be used to illustrate this, but one of the very best draws on the experience of Oregon Steel Mills, Inc. (OSM).

OSM is substantially smaller than Nucor, has a narrow product mix, primarily plate and pipe, and has not been as innovative as Nucor in terms of its use of technology. OSM does use a unique casting system, but it has invested in mainly proven technologies to achieve a low cost, high quality position in its chosen markets. While these contrasts with Nucor are real and significant, there is no contrast between the two firms when it comes to success in the marketplace. OSM has reduced the labor hours required to produce a ton of steel by more than half since the mid–1980s. Its current level of about two and a half is on par with the best globally in its main product lines. As a result OSM competes successfully for orders against the most efficient plate-producing companies in Europe and Japan, and in 1993 it won a major order from Thailand's petroleum authority to supply the pipe for a deep undersea link between an offshore gas field and Thailand's national distribution system. Its growth has been impressive: from 1986 to 1994, its sales increased by a factor of ten, tonnage sold rose by a factor of eight, and its financial performance was superior to that of all the integrated steel companies in North America.[12]

Tom Boklund is Chairman and CEO of Oregon Steel Mills. When it comes to describing his successful management philosophy, the first thing on Boklund's mind is breaking down the barriers between management and labor. In fact, there is an open atmosphere at OSM that encourages extensive communication and cooperation, but it came at great cost—a long hard strike in 1983–84.

At that time the company's labor force was unionized, and OSM was seeking a 17 percent wage reduction and a number of other important contract changes, including replacement of seniority with merit as the criterion for promotion. The wage concessions on the table were large, but so was the competitive and financial crisis that OSM faced. The union struck, management operated the mill and hired replacement workers. About half of those hired during the strike were union workers who crossed the picket line. As a result the strike failed, and ultimately the workers at OSM voted to decertify the union.[13]

The workers hired during the strike were generally new to the steel industry. For them, there was no burden of a history of adversarial relations with the firm to "unlearn" and the former union workers who remained were receptive to change, as it had become obvious that change was necessary. However, to capitalize on this opportunity, swift and decisive action was required.[14]

Right after the strike, Boklund moved quickly to eliminate all policies that discriminated between the employees in management positions and the plant workers, unless there were obvious, valid business reasons to retain them. Time clocks were eliminated and all employees were changed to salaried status. Executive parking spaces were eliminated, and the executive lunch room was closed. Company cars were limited strictly to outside sales personnel.[15]

Of all these changes, the elimination of hourly pay was most contentious, and the resistance came from the supervisors. Some simply did not believe that employees who were used to an hourly pay scheme could be trusted to work to their full potential without the close monitoring of behavior that comes with punching a clock. They were certain that the new flexibility that Boklund was proposing would backfire, but they were wrong. Workers on the plant floor welcomed the opportunity and recognized the importance of the trust that had been placed in them. Absenteeism, which had averaged almost 3 percent before that action, fell to less than 2 percent.[16]

Boklund's next step was to introduce an employee stock option plan (ESOP) and soon after that he introduced profit sharing. With these measures he cemented the commitment of workers to the firm. By the end of 1984, the ESOP had purchased a majority of the outstanding stock, and the remaining stock (excluding 10 percent held by management) was purchased by the company and retired. The employees became part of the ownership, and their stake grows annually as the company contributes to the ESOP every year.[17]

OSM's profit sharing plan played a supportive role in aligning the interests of the employees with that of the company. The plan initially paid 35 percent of pretax profits into a pool from which quarterly payments were made to all employees. This was reduced to 20 percent in 1988 in order to bring it more in line with publicly traded companies. Payments to workers under the plan have been made each year since 1986, but the payments have varied from year to year as profits have fluctuated. In recent years, profit sharing for production workers has ranged from 5 to 30 percent of base salary. While considerably less than Nucor's production bonuses, management feels that relating employees' compensation to company profitability supports the concept of fairness, which is fundamental to the company's culture.[18]

The large fluctuations in profit sharing, many of which reflect changes in the company's product mix and market price, place a substantial responsibility on management to adequately communicate the reasons for the change. If the employees do not have a good grasp of these issues and profit sharing declines while production remains high, employees are likely to become disillusioned and trust may be eroded. Constant communication is critical here, as it has been shown to be in other respects as well.

Accountability and Reward for Performance

Trust and commitment are important, but so is the responsibility that goes along with trust. People have to be held accountable for their actions before they can take pride in their accomplishments. In companies like Nucor and Oregon Steel Mills top management has, in effect, given employees control over their work space—whether that space be measured for a worker as his or her job in the mill, for a manager as control of the melt shop, or for the director of operations as responsibility for the entire plant. Correnti's attitude about resisting the urge to wrestle control from a subordinate is based on the importance he and Nucor attach to these principles. In his words:

> I don't want to take the steering wheel out of the melt shop manager's hands with respect to purchasing or any other operating decision. Because then if his performance is not adequate, he will say, "Correnti bought the electrodes, he paid too much or he bought the wrong quality." I want to keep the decision making in his hands so he can look in the mirror at the end of the day and say, "I made money today for Nucor because I made the decision" or "I lost money today for Nucor because I made the decision,"

> not somebody back in Charlotte, North Carolina. You want to put the decision making as close to the source as possible. You want to keep pushing it down, to let the operators control their own destiny.[19]

When employees have this kind of control, accountability for performance is easy to establish. The most successful minimills make use of the linkage between responsibility and accountability to establish systems that reward good performance and reinforce the concept that what an employee does directly affects not only the company, but the employee as well. The concepts of ownership over the job and being rewarded for the way that job is carried out are important elements in establishing a culture in which all employees internalize the company's goals in their approach to their jobs. In the most successful minimills this linkage has been used very effectively.

Nucor is widely recognized as among the industry leaders in using performance-driven compensation. For example, the base pay rate for nonmanagerial employees at the Hickman plant is about $10 per hour, roughly half of what a unionized steelworker makes in an integrated facility. Production incentives, however, fill that gap, and they have consistently elevated average take-home pay at the mill to more than $50,000, which is above the norm elsewhere in the industry.[20]

All Nucor employees are covered under one of four compensation plans, each with incentives related to meeting specific targets. Employees involved in manufacturing are paid weekly bonuses on the basis of the production of their work group. These bonuses are based upon the amount by which actual results exceed the preset standard for the operation. The plan creates peer pressure for each individual to perform well. No bonus is paid if equipment is not operating, thus creating a strong incentive to keep the equipment in top operating condition. Maintenance personnel and production supervisors are part of the same bonus group and receive the same bonus as the people they supervise. Production incentive bonuses average between 80 and 150 percent of the base wage.[21]

Nucor department managers earn incentive bonuses paid annually based on return on assets of their facility. Nucor pays no discretionary bonuses. The bonuses can go as high as 80 percent of base wage. The incentive plan for nonproduction and nondepartment managers is based on return on assets for each facility. Participants include accountants, engineers, secretaries, clerks, receptionists, and a number of other employee classifications. A significant part of each senior officer's compensation is based upon Nucor's return on stockholder equity, above a specified minimum return. The maximum is 400 percent of base pay, and in a good year it will average about 300 percent.[22]

Another excellent example of a firm that recognizes the importance of decentralization and capitalizes on this with an explicit incentive pay scheme is Birmingham Steel Corporation. This highly successful minimill was formed in 1984 to acquire the existing plants of a firm that manufactured mine roof support systems. In the last ten years, its growth has been driven by acquiring steelmaking facilities from a number of companies, consolidating assets, and restructuring manufacturing processes into highly efficient, profitable operations. It has been profitable since its inception, its financial performance has been high by industry standards, and its sales have grown by almost eight times since it was formed.[23]

James Todd, Chairman and CEO, is widely recognized as the driving force behind this success. He describes his management philosophy succinctly: "I believe in hard work, paying people well for good effort and holding people accountable for their decisions. The key to attaining superior performance is to motivate people to expend high levels of effort to accomplish your goals. Incentive systems are a major part of this, but so is getting to know your employees, understanding their concerns, and securing their trust."[24] He carries through with this in practice.

All Birmingham employees are paid an incentive. The hourly workers are paid on the basis of good tons produced for each week compared to a set standard. Each product has its own standard. Management is paid out of a pool which is 3.5 percent of pretax profits. Tonnage was selected as the basis for the workers' plan because it is easy to understand and cannot be manipulated for one accounting reason or another.[25]

The production workers are paid a base hourly wage and in addition receive incentive payments. These vary by plant. For the company as a whole, during the past few years incentive payments have amounted to about 50 percent of base wages. The incentive payment is computed weekly on the basis of actual good quality tons produced compared to a production standard, and not on the basis of profits as is done at Oregon Steel. The amount of the incentive is put into a pool. All the man-hours worked during the period by those eligible are divided into the pool, and each employee receives the same amount for each hour worked. The employees are paid weekly and the incentive payment is in a separate check. The keys to the incentive program are that it is easy to understand, payments are frequent, and it is related to the quality of what is produced.[26]

Production-based bonuses like those at Nucor and Birmingham Steel are not necessary for success, but the linkage between personal responsibility and accountability upon which production bonuses depend is common and regarded as critical in the very best minimills. Oregon Steel Mills does not have a production-based incentive system; rather, as we have seen, this firm uses the firm's profits, not tonnage, to tie "performance" to financial reward. There is a second key element in this linkage for Oregon Steel, however, and that is individual performance appraisals. These appraisals are conducted semiannually for all employees by their supervisors. Performance objectives are established and each employee is assessed by criteria to which he/she has agreed. While very time-consuming, the performance appraisals are a big part of making sure that each employee knows what he/she is being measured against. They are powerful motivating tools because salary increases and promotions are based upon how well the employee performs against these objectives and standards.

Integrated producers and unionized minimills are limited by contract conditions in the extent to which these principles can be applied. When seniority replaces accomplishment as a criterion for reward, the benefits of decentralization simply are not there to the same extent. Graham saw this at USS and Armco, but he applied the very same principles that we find at Nucor, Oregon Steel, and Birmingham Steel as broadly as possible to the ranks of management. In Big Steel or the minimills, there seems to be broad acceptance of the fact that effective leadership depends on establishing a culture of empowerment and capitalizing on it.

Risk Taking

Perhaps the most difficult aspect of empowerment for top management is the balance that must be struck in a firm between risk and reward, and firms, like people, run the gamut from being risk averse to aggressive risk takers. Nucor sets the standard in the steel industry for firms that thrive on the challenge that comes with confronting risk. As we have seen, this firm has found great opportunity in the implementation of untried technology, especially by its entry into thin-slab casting, and it has won on a grand scale because of its success in doing so. Less well known, perhaps, is the fact that Keith Busse, who was responsible for the Crawfordsville project, had started up two other operations for the company, a bolt factory and a factory to produce prefabricated metal buildings, neither of which was a financial success! The bolt business ran into stiff price competition from Canadian producers, and the prefab building business encountered quality problems as well as severe competition. Despite the lack of success, Busse had demonstrated a good understanding of technology and the drive required to start businesses from scratch. By appointing Busse, Iverson, showed that lack of success, for reasons beyond one's control, would not be penalized and that risk taking would be rewarded. Expressions of confidence of that magnitude go a long way toward imbuing a company with a risk-taking mentality.[27]

Another aspect of the risk-taking culture at Nucor is reflected in the decision to give Mark Millett the responsibility for the melt shop at Crawfordsville. He was twenty-eight years old, with a bachelor's degree in metallurgy, and little steelmaking experience prior to his employment at Nucor. But management was not afraid to place significant responsibility in the hands of a person who it believed had the technical and entrepreneurial skills to solve the difficult and unpredictable problems that would arise in a new start-up. Thus tenacity and ingenuity were valued more than a proven track record. Millett demonstrated the wisdom of this approach over and over again as he, Busse, and others successfully brought the thin-slab casting process to a commercial success.[28]

The trust placed in Busse and Millett demonstrates Nucor's culture. Aggressive, innovative individuals were given the lead. Experimentation was rewarded. Failure was accepted as a part of reality, and advancement did not depend on the absence of failure. If it had, risk taking would certainly have been discouraged.

Nucor attempts to lessen its risks by working closely with the vendors supplying the new technology. Their joint venture with Yamato is a case in point. The Japanese partner had the technology and the experience that Nucor needed. Nucor sent twenty of the new Arkansas hires at a time over to Japan to get experience on Yamato's caster. Each group was committed for two- to three-week stints. When they returned they were supported by Yamato's engineers. But while there were as many as twenty Japanese engineers at the outset in the U.S. operation, that number fell to just two or three very quickly. Nucor employees learned by doing, and made improvements to the process as they went along. Today they cast wider and heavier than Yamato had ever envisioned for the process. As a result, the market potential for Nucor in heavy structurals opened nicely, and their competitors' market shares plummeted.[29]

Nucor took the same approach later when it committed to the SMS thin-slab casting process. Again, Nucor sent a number of the workers who would be involved in running the new plant to Germany to be trained at the foundry and to become familiar with the new equipment before it was shipped to Crawfordsville. The Nucor employees who would operate the new mill were intimately involved in its construction phase, including the installation of the equipment. They were also responsible for the start-up—very few German engineers were involved. Learning by doing in this way helps Nucor to build strong team spirit. All the workers are involved in the successes and failures of the start-up phase and assume "ownership" of the process and the equipment as a result.[30]

While Nucor sets the standard, it is by no means the only successful steelmaker who understands that big rewards are rarely associated with small risk. In fact, improvements in terms of cost control or quality on a day-to-day basis often depend on sustaining the kind of risk-taking culture that one observes at Nucor, if only in less visible form. Oregon Steel is not known as a pacesetter in technology, but it has pushed proven technology to the limit in order to achieve its cost and quality goals. One key to its success in doing this is revealed by Kevin Ratliff, Manager of Operations at OSM's Portland plant. As he put it,

> Productivity problems in many of the integrated companies are related to the fact that people won't give the company more and that managers won't push their workers. Past practices are a major impediment to change and this is reinforced by labor contracts that have taken away from management many management prerogatives. This is not the situation here. Our managers think they can ask people to do anything as long as it is reasonable, and they expect their workers to respond. *Nobody is penalized for failure in trying something new. We expect our managers and workers to try new things.*[31]

Decisiveness and independence are valued, and these attributes are an inherent part of a risk-taking culture. This firm recognizes that there are limits to the benefits that accrue from consensus building—limits that are reached when the process itself becomes a substitute for action.

Continuous Improvement

If a modern steel firm is going to survive and prosper, it must be cost- and quality-oriented; in fact, producing to customer requirements is the price of entry into the marketplace. For lower value-added, basically commodity products, there is not much any of these companies can do to gain a competitive edge on the product side other than to ensure that its quality and on-time delivery performance equal or exceed that of the competition. Therefore the focus is on the cost side, where a company can achieve an edge in the marketplace if its costs are lower than its competitors', thereby giving it greater latitude in pricing. James Todd at Birmingham put it this way: "We realize that we have little control over the price of scrap or the market price of steel, but we do have control over our production processes. Thus, the way for us to make money is to continuously improve our processes and to drive down our conversion costs. We have an objective of being able to convert scrap into steel for less than $100 a ton."[32]

The drive to be the "low cost" producer in a market, to set the world standard, is pervasive in the industry. One hears it over and over again on plants visits whether talking to top management or shop supervisors, and whether the plant involved is in Alabama, Indiana, or Oregon. For that matter, the most successful U.S. minimills will not enter a market unless they are convinced that they can set the standard for low-cost production. Once there they intend to keep the market by continuous productivity and quality improvements.

This attitude—never being satisfied with the *status quo*—is the norm for managers of these minimills, but it goes well beyond process improvements. Focus on process advantage has a direct counterpart in focus on market advantage. Their investments are strategic and aggressive in terms of penetrating new markets and expanding existing market shares. Here too the emphasis is on continuous improvement, but the goal is market position.

We saw this first at Nucor where the strategy is typified by the company's success in exploiting technological advantage in order to seek new markets. Other successful minimills may not pursue these advantages through technology to the same extent as Nucor, but they are equally dedicated to a continuous reassessment of market position and strategic investment to improve that position.

Oregon Steel Mills invested more than $200 million during the mid–1980s through the early 1990s to acquire assets to expand its product line and to move into more sophisticated steelmaking techniques to upgrade its product mix.[33] And in 1993 it completed the purchase of the CF&I Steel Company to further its diversification into new product lines. To ensure a competitive edge, it sold a 10 percent equity stake in CF&I to Nippon Steel and Nippon agreed to transfer technology to the company to enable it to produce rails that are harder and longer-lasting than the competitors'.[34]

Birmingham Steel's basic strategy was similar in many respects to OSM's. It was formed in 1984 to purchase a company that operated two old minimills and a group of facilities for manufacturing roof support systems for the coal mining industry. Subsequently it purchased five other minimills, most of which were inefficient facilities using out-of-date technology. For these, the technology was upgraded and approximately $300 million was invested to increase productivity and lower manufacturing costs. In 1993, it acquired the American Steel and Wire Company (ASW) as part of its strategy to diversify its product mix into more sophisticated, higher value-added products. The acquisition reduced the company's reliance on the lower value-added commodity rebar market with these products falling from 60 percent of sales prior to the acquisition to 40 percent afterward. Shortly after purchasing ASW, the company entered into an agreement with Sumitomo Metal Industries of Japan for technical assistance for the design and start-up of a new bar and rod mill for ASW and for upgrading an existing rod mill. This transfer of proprietary knowledge and technology to ASW will enable it to significantly upgrade its capabilities, another example of the company's ongoing strategy for improvement.[35]

Successful leaders in the steel industry know that making good in this kind of drive is not accomplished by publishing a manifesto; nor is it a matter of announcing the company's "program of the year" and giving it a cute name. Rather, as Todd suggests, the drive has to be continuous and permeate the firm, and it has to flow naturally from a culture in the firm that nurtures and sustains it. From the experience of this industry, the culture we are talking about seems to have a few central elements: (1) a sense of purpose must be ingrained, (2) trust has to be established, and this goes both ways, from management to workers and from workers to management, (3) a firm must take maximum advantage of opportunities to decentralize decision making, (4) risk taking must be encouraged, and (5) there has to be accountability and reward for performance.

In the most successful steel firms the work environment is defined by these characteristics, and the importance of striving to improve is widely acknowledged. Tom Boklund at Oregon Steel might just as well be speaking for Nucor's Iverson or Birmingham's Todd when he says,

> Creating an environment in which people are encouraged to change and to constantly improve is a priority. Failure of new ideas is not punished, people are encouraged to grow, and benchmarking of excellent companies is carried out. Our approach to business is to run our operation in an 'optimally effective' manner. This means that we get the most out of each of our operations. But for this to occur, management must provide the leadership."[36]

For that matter, Boklund might be speaking for the industry as a whole. The conditions for success in this respect really do seem to emanate from the top.

Summing Up

Achieving and sustaining competitive advantage in the marketplace require strong leadership, and strong leaders are not people who micromanage a firm. They create an environment or culture that draws the very best out of the people who comprise a firm, and they motivate those people to strive for goals that help a firm to succeed. It is the role model established by top management that sets the stage for the company's ability to compete, and it is the policies and practices that evolve that creates the "culture" of the workplace. This culture gives rise to motivation levels and behaviors that determine the relative competitiveness of the company.

The next chapter delves into the notion of a "high-performing" workplace. We shall find that the elements of effective leadership that we have exposed are inextricably bound to the objective of gaining maximum productivity advantage in technology and human resource practices.

6

High-Performing Workplaces

The culture of success that we have described cannot be fully understood or appreciated without learning more about the ways in which the technology of steelmaking and the people who make steel interact with one another to shape competitive advantage. The relationships involved are highly complex. Hence any attempt to isolate one factor among many as the "key" to success in the marketplace and high rates of profit would certainly be misplaced. In fact, it is precisely in the interactions among elements of competitiveness that one must look for insight on American steel's rebirth.

To achieve success, managers have to understand and take advantage of the interlocking nature of investments in their firms. The first step in this awareness is learning to define the concept of investment broadly. The objective must be to make the most out of investments of all kinds—investments in machines, in the skills of employees, in human resource practices, and in the way that work is organized. All these are interwoven. If investment in any one is lacking or wrongheaded, those in the others cannot rise to their full potential. In this chapter, we explain this critical aspect of competitiveness. In the process we will learn why the culture of a firm is fundamental to its performance.

Gains in Productivity and Quality

The productivity gains of integrated producers and minimills are widely acknowledged. Less well understood, perhaps, have been the parallel and related gains in steel quality. As few as ten years ago, American steelmakers were being forced out of many domestic markets by foreign producers who were much better able to meet customer demands for quality steel delivered in a timely way. The productivity and quality gains were part and parcel of one process. Although foreign steel producers still account for a significant share of total U.S. steel consumption, their market penetration no longer reflects cost or quality advantages over U.S. producers to a significant degree. Rather, American steelmakers are currently producing near to their capacity, and imported steel has helped to meet demand that would otherwise go unsatisfied at existing prices. Indeed, a significant portion of today's imports are in the form of semifinished slabs.

Along with strategic investments by American firms to secure particular product markets came the installation of ladle refining, vacuum degassers, and continuous casters that allowed companies to produce high-quality steels with fewer man-hours. Today, nearly 90 percent of American-made steel is continuously cast.[1] The importance of ladle refining in modern steelmaking is suggested by the fact that fewer than five ladle refining furnaces existed in the United States prior to 1984, while more than fifty are in operation today. Similarly, ten years ago vacuum degassing was used only in the production of specialty steel; now all major integrated companies have such equipment in at least some of their plants.

Both these processes, ladle refining and vacuum degassing, can be thought of as steps that bring about final properties in steel so that steel products meet the high standards of today's end users. In effect, they give steelmakers the ability to "fine-tune" the chemistry of steel in ways that enhance characteristics like surface quality. Whereas the manipulation of steel chemistry had taken place in the huge BOF steel furnaces just a few years ago, today hot steel comes from those furnaces directly to a ladle where temperature and chemistry can be controlled much more precisely. Further refinement in vacuum degassers has had much the same effect: By moving hot steel directly from the BOF furnace to an environment where the material's properties could be carefully controlled, immense quality improvements were realized and new steel products were developed.

Benchmarking Technology and Efficiency

The significance of these advances spills over in important ways to the human dimensions of competitiveness. To appreciate how and why this is so, it is helpful to talk about standards of performance that can serve as benchmarks of industry and firm progress. One foundational study of this sort was carried out in 1992–93 by scientists and engineers at Carnegie Mellon University (CMU).[2] Detailed questionnaires were completed for sixteen integrated plants representing firms in the United States, Europe, Japan, and several smaller steel producing nations. Data also were obtained by questionnaire from twenty EAF producers. This group included U.S. and European minimills as well as some electric furnace plants owned by foreign integrated firms.

For each plant, data were obtained on the technologies used and the efficiency of production along certain specific dimensions. On the technology side, the focus was on the implementation of critical processes like sublances in oxygen steelmaking, ladle furnaces, and coal injection in the blast furnaces. All in all, technical information of this sort was gathered for each stage of steelmaking. On the efficiency side, they looked at about fifteen measures such as the rate of consumption of fuel (e.g., coke, electricity), tap-to-tap times in furnaces, oxygen use, and electrode consumption—depending on the type of furnace being examined. Importantly, the data collection and analysis were subjected to close examination by industry representatives. Their feedback on the study design led to further data gathering and to refinement of the indexes that were developed in the study to measure overall technology implementation and the efficiency of manufacturing in each plant.

The findings from this study for the sample of integrated plants are displayed in Figure 6.1. The measured indexes for each plant are plotted in this figure, and the scatter diagram that results suggests the expected correlation: Higher rates of implementation for critical technologies are

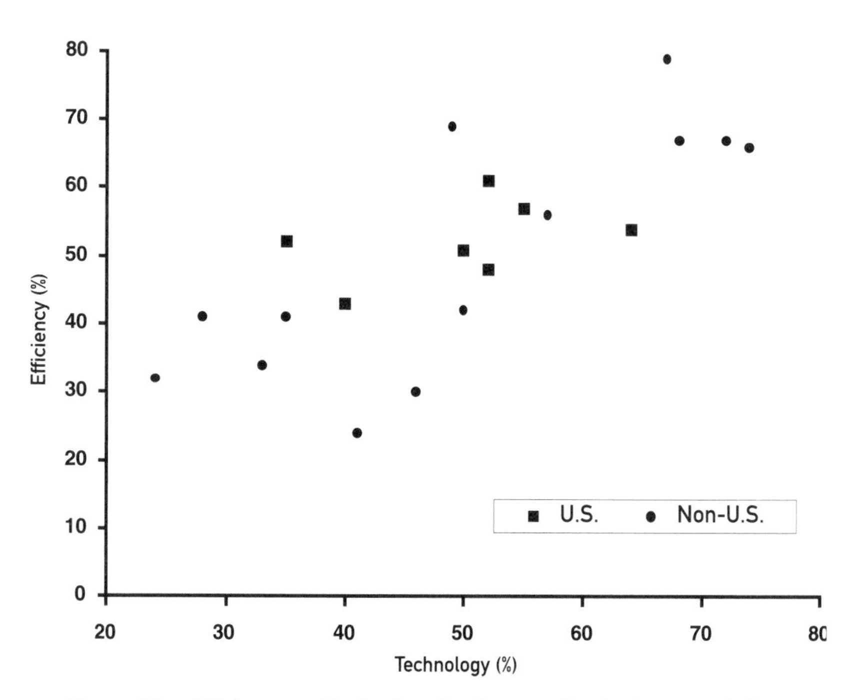

Figure 6.1. Efficiency and technology implementation for integrated plants.
Source: R. J. Fruehan, L. Brinkmeyer, R. Dippenaar, and Y. Zhang. "Manufacturing and Technology Assessment of International Steel Plants." *Iron and Steelmaker* (January 1994): 29.

associated with higher rates of efficiency. It is also evident that many of the U.S. plants in the sample compare favorably to their foreign competitors.

As is often the case, however, real insight is gained not from the general trend in these numbers but from the wide deviations in measured efficiency that can be found in plants with similar rates of adoption in critical technologies. Consider the four U.S. plants clustered in the 50 to 55 percent range for technology implementation. The index for efficiency in these plants ranges from 48 to a high of 61, indicating that from the worst performer to the best one there is an advantage in terms of efficiency of about 35 percent—and this for plants with approximately the same rate of technology implementation. Even greater variance can be observed in efficiency comparisons for some of the foreign plants.

Getting at the factors that lie behind such differences in a scientifically valid way is extremely difficult. The sample involved is small, and the explanations that can be offered are many. In such situations there is no substitute for detailed knowledge of the plants themselves and knowledge of the skills that people in those plants bring to bear on its operation—whether the skills involved are technical or managerial.

In explaining such differences the study team pointed to two factors as being especially important. First, having technology in place and knowing how to get the most out of it are very distinct things. The integrated plants with the best performance records for a given level of technology implementation were those with the most capable production workers. There is no substitute for technical know-how and the high performers seek this aggressively, even if that means purchasing it directly from foreign competitors or securing joint agreements with them. Examples of this abound in the United States: LTV Steel with Sumitomo Metal Industries, Inland Steel with Nippon Steel, National with NKK, and Armco Inc. with Kawasaki.[3] In these and other cases, major producers have aligned themselves with firms that have demonstrated ability to make optimum use of technology, and they have gained in efficiency as a result.

But some of the advantage exposed in Figure 6.1 has little to do with technical know-how per se. The CMU study team points to a second major factor in this respect which shows up for firms of similar technology in the ways in which various stages of production in integrated plants come to work together in defining the plant's production process as a whole. As we explained in earlier chapters, integrated plants can be described as a series of processes, and all processes must be compatible and coordinated in order to achieve maximum efficiency and quality of product. A plant that "overinvests" in one operating unit and neglects another will create a bottleneck in the production process that translates directly into diminished efficiency and operating cost disadvantage. Poor coordination of this kind means that a plant can be pressing against capacity constraints in one part of its operations while other parts are idled or operating far below capacity. So in a very real sense the overall productivity of a plant or the quality of its product is determined by the least efficient unit operation or the one that causes quality problems.

The study team's findings tell a somewhat similar story for the EAF shops that were surveyed, and Figure 6.2 displays these results. In this case, however, the study team points to operating know-how as the dominant factor in explaining the efficiency differences, for the simple reason that EAF production is not complex and imbalances are more easily avoided. But even this understates the human dimension of competitive advantage, for the competitive gains won by minimills in the United States go well beyond the ability of a few firms to eke out greater productivity and quality from comparable technology.

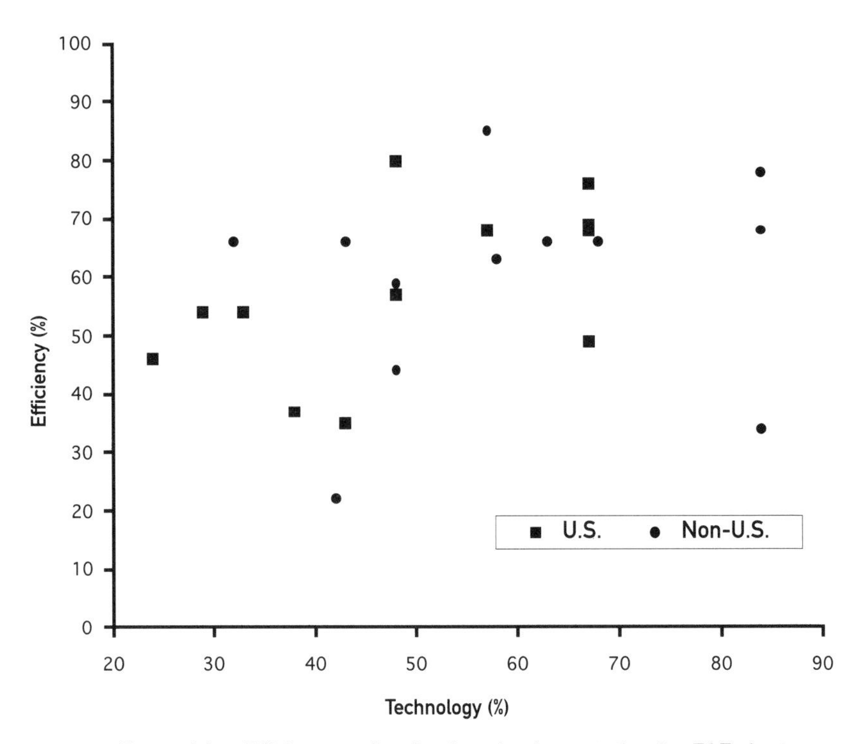

Figure 6.2. Efficiency and technology implementation for EAF plants.
Source: R. J. Fruehan, L. Brinkmeyer, R. Dippenaar, and Y. Zhang. "Manufacturing and Technology Assesment of International Steel Plants." *Iron and Steelmaker* (January 1994): 26.

Change in Technology and Change in the Workplace

Consider the impact of continuous casting on the work environment of a steel mill. It is easiest to visualize this in an integrated mill by comparing activity before and after the installation of the caster, but the implications generalize to minimills as well. Steelmaking in an integrated facility without a continuous caster relied on inventories of solid steel between the BOF steel furnace and a plant's rolling operation. Molten steel would come out of the BOF to be poured in ingot molds, where it stood until it had solidified. Then it would be stored to await the next stage, which meant committing the steel to a semifinished shape—billets or blooms for long products and slabs for flat products. Rolling operations followed. Of course, reheating was necessary to get from the ingot to the billet, bloom, or slab, and again from one of those forms to the final rolling operation.

Continuous casting changed this process radically. Every caster is built for one and only one semifinished shape. A mill commits to a market, and the caster is dedicated to that market. Competing in long products means purchasing a billet caster or a bloom caster; competing in flat products requires a slab caster, with slabs of traditional thickness in integrated plants or "thin" slabs in minimills. Once that decision is made and a continuous caster is installed, steel coming out of a BOF furnace is never allowed to solidify before the casting operation; it goes directly from molten steel out of the furnace into one of the three basic semifinished shapes by passing through the continuous caster.

The advantage of the traditional process that relies on ingots was closely related to the fact that integrated steel mills used to produce a wide variety of products. With ingots in inventory, raw materials (steel ingots) could be directed to any one of the lines that were in place to produce semifinished forms. The choice depended only on the order wait list; for example, if reinforcing bar had to be made, the ingots moved to the billet mill. Also, while steelmaking and steel rolling operations were mutually dependent, the inventory of ingots provided a buffer between them that insulated each from interruptions that might originate in the other. The BOF could continue to produce steel for ingots when the casting operations were jammed up or idled, and the casting operations could obtain ingots from inventory if problems arose in the steelmaking shop.

The cost savings with continuous casting, however, were demonstrable. The single largest source of savings was in the "yield" or usable steel that one could obtain from a given heat in the BOF furnace. In casting ingots, fully 10 to 20 percent of the steel that came out of a BOF furnace was lost to immediate use. The wastage had to be gathered up and recycled as part of a subsequent furnace charge. There were also substantial energy savings, of course. Ingots are huge, and the energy required to reheat them and roll (squeeze) them into the semifinished shapes is very substantial. Interestingly, the capital cost differential between the two processes was not terribly significant. The gain associated with eliminating the old casting operations and inventories of ingots was offset by the cost of the new casting operations.

At least as important as these changes, however, is the change that was forced in the organization of work in a mill. With ingots gone and their inventories eliminated, steelmaking becomes a continuous process. For the integrated mills, this means that hot metal (i.e., pig iron) from the blast furnace must move to a BOF steel furnace and then on to a caster, and there are no inventories to serve as slack. For the minimills the story is much the same—scrap to billets (or blooms or slabs) in a continuous operation.

The profound importance of changes in manufacturing operations of this kind were first brought to general attention by the work of Womack, Jones, and Roos.[4] Their book, *The Machine that Changed the World*, explains how changes in the process of manufacturing automobiles translated into competitive advantage for Japanese and, later, American car manufacturers. Their analysis leads to the concept of "lean manufacturing," which embodies the idea that a high degree of coordination and cooperation in the workplace can dramatically lower production costs and improve quality. At the heart of this revolution in organization was the elimination of inventories. Without the crutch of relying on these buffers, slack in the workplace must be eliminated. People and machines have to perform well, they have to perform together, and they have to produce a good product, or everything fails. The Japanese were first with just-in-time operations, but now the concepts involved have been generalized and extended to automobile producers worldwide and to manufacturing operations of all kinds.

Steelmaking has been affected dramatically by a somewhat similar change in the production environment. If a plant manager expects to move tons and tons of hot steel around in a continuous way from one stage of operations to the next, process control and coordination are essential. Mistakes can be disastrous. Indicative of the danger is this simple rule of caution on the shop floor of a mill: If water spills on molten steel, you get steam. If molten steel spills on water, you get an explosion!

There are side benefits to paying this kind of attention to the manufacturing process. For one thing, capital utilization rates are likely to improve. If the parts of the process have to be coordinated and organized, they have to be made to fit. There is strong incentive to eliminate excess capacity in any one part so as to balance the whole. Product quality will rise too. It is no coincidence that steelmakers invested heavily in ladle furnaces and vacuum degassers at the same time that they were investing heavily in continuous casters. The casters made it necessary to learn to work with hot steel and keep it flowing. It is a small matter to extend that control to intermediate stops at a ladle or vacuum degasser for adjustments in chemistry before proceeding to the casting machine. Success in coordinating activity breeds further success.

Complements in Production

One way of thinking about the relationship between continuous casters and ladle furnaces (or vacuum degassers) is to recognize that an investment in the caster made further investments (in ladle furnaces, for example) more valuable. Taken together the investments allowed the firm to move into higher value-added products, and their rates of return were enhanced as a result. This is a classic description of "complements" in production as defined by economists, and by "classic" we mean that the idea has been around for a very long time, say a hundred years. Investment in either one of the two complements *increases* the rate of return that can be expected from investment in the other.

This is a very powerful concept, precisely because it generalizes widely. When the scientists at CMU pointed to technical know-how as the determining factor that explains efficiency gains for plants of similar technology implementation, they were reacting to the fact that firms can decide to invest in "human" capital just as they can decide to invest in machines or "physical" capital. In both instances a conscious decision is taken to use a firm's scarce resources to build something that will pay off for a very long time. In one case, a knowledge base is being built; in another, it is a continuous caster or ladle furnace. But the analogy is direct and clear. When American firms went to known performers like Nippon Steel to learn how to get the most out of new or existing technology, they were making an investment.

The analogy of human and physical capital can be used to extend the idea of complementarity. Once the investment is made in a continuous caster, vacuum degasser, or ladle furnace, the rate of return to investment in human capital increases dramatically. Investments in some technologies permit a "dumbing down" of the workforce in a company, but investments in these technologies have had just the opposite effect. Before the advent of ladle metallurgy, for example, fine-tuning the chemistry of steel really meant guesswork in adding needed alloys. Now, a worker in a control room high above the furnace is keeping track of half a dozen display screens, checking tolerances of the chemistry, and making changes as required. In this case and others like it in the mills, new technology and human capital must be developed in use at the same time. What we recognize in the marketplace as high-performing firms are the very firms that have taken full advantage of that kind of complementarity. And the potential range of these advantages is very wide indeed.

The importance of complementary relations such as those we have described in understanding today's workplace was explained by Milgrom and Roberts.[5] Writing at about the same time as Womack, Jones, and Roos who were studying the automotive industry, Milgrom and Roberts realized that complementarities could help explain why similar kinds of change in the organization of production within firms appeared to be so widespread and cut across so many different industries. So pervasive was this phenomenon that they coined it "modern manufacturing." The term was applied when the introduction of new technology widened the range of products in a plant by introducing flexible machines that reduced the setup costs associated with changing from one product to the next. In automobile production, for example, they refer to modern machines that can be set up by engineers over a weekend to change from one model to the next, as compared to design changes of the same sort that might have meant a plant shutdown of several weeks in the not to distant past. Once that kind of technology is in place, a whole host of other changes in the organization of production become more feasible and more profitable—changes ranging from the elimination of inventories to the introduction of teamwork.[6]

Advances in steelmaking technology, however, have encouraged greater specialization in product lines for any given plant, at least in the sense that the choice of casters limits the range of final products a mill can produce. Greater flexibility and smoother changeovers between product runs have come in rolling operations, but one cannot point to major technical innovations here of the sort that stimulated the interest of Milgrom and Roberts. Nevertheless, their fundamental point is applicable to the steel industry and provides a unifying framework for much of what we have observed: If one wants to explain the basis of competitive success, one must understand the ways in which investment breeds investment, and change complements change.

Human Resource and Organization Practices

We have already seen one way in which complementarity can link capital and labor. But the decision to invest in human capital is only the first of many steps by which investment in steel technology has changed the work environment in steel mills. Continuous casting also places a premium on the *flexibility* of labor. As hot steel moves from the BOF furnace to the caster, the ladle used to fine-tune the chemistry of the metal can also serve as a way station. If a problem arises in the caster, for example, and a heat of steel has to be held up, the furnace at the ladle can keep the steel active while the problem is corrected. This kind of delay cannot be indefinite, however, and it is costly. The sooner the problem causing a holdup can be resolved, the more efficient will be steelmaking. Looked at in this way, the introduction of a continuous caster establishes a "continuous" process of steelmaking as the ideal. As firms strive for that ideal, costs will fall because less energy is being used per ton of output and more metal is being made with a given set of machines. In short, productivity is enhanced.

Decentralization, Again

Once committed to continuous casting, it is critical to resolve problems quickly. Better yet, they should be prevented from occurring in the first place. To achieve that end the people closest to the machines and closest to the problems must have the know-how to act effectively before a crisis arises and the authority and ability to correct problems when they do arise. Then, they have to act quickly. In order to make effective use of people on the plant floor and take maximum advantage of the information they have about the production process, decision making must be decentralized. As we have seen, that rule is at the core of high-performing steel mills.

Typical of the importance attached to these practices in achieving a competitive edge are the views of Tom Boklund at Oregon Steel Mills: "Pushing decision making down as far as possible within the organization is an ongoing objective. . . . The theory is that the fewer the people and the more responsibilities each has, the greater will be the incentives to make decisions fast and to establish processes for making decisions at lower levels."[7]

As we demonstrated in Chapter 5, decentralization and empowered decision making have been taken the farthest by the minimills—especially by Nucor—but these objectives have been important for integrated producers like U.S. Steel too. The main difference is that firms like Nucor were able to build much of their corporate structure anew, while the integrated companies were trying to change an existing structure and an existing culture.

It is worth following the restructuring at USS to observe how changes in organizational structure can drive changes in decision making. As USS shuttered plants and laid off tens of thousands of employees, it cut out at least three layers of management, eliminated all "assistant" and "assistant to" positions and broadened the responsibilities of those who remained. For instance, a works manager who in the early 1980s was responsible primarily for daily operating decisions, now became directly responsible for quality assurance and production planning. (The quality assurance and production planning people in the plant previously reported to managers at corporate headquarters.) These changes, coupled with a greater focus on meeting the needs of customers, pushed the works managers in the direction of greater customer involvement.

As the responsibilities of the works manager increased, decision making within the plants was driven downward to area managers and eventually to the plant floor. Each manager had greater responsibility and fewer staff, and thus had to rely more on those closer to the actual operations to make decisions. Without such reliance performance would suffer. Increased responsibility like this makes it feasible to hold each level accountable for results as their greater span of control gives them the ability to manage the factors that affect their performance.[8]

Since the reorganization, responsibilities retained in USS's central office are those that affect the corporation as a whole. For example, capital investment decisions above a certain level and the allocation of capital are held centrally. Investment funds are scarce resources and have to be allocated in light of a company's overall strategic priorities and not those of an individual plant. Similarly, responsibility for policy on labor relations is retained by the central office in order to ensure uniformity in the way labor contracts are interpreted and implemented across plants.

Throughout the USS organization, however, there is a strong belief that decisions should be make by those closest to the problem, and that those people should also be in a position to implement the required solutions. Management has recognized that operations, quality, and customer service decisions can be made faster by more knowledgeable individuals if those closest to the problems are empowered to make the decisions that affect the performance of their processes. On the plant floor, this means that the operating crews are given objectives to meet in terms of tonnage and quality, but they have great discretion in how these are to be accomplished as long as standard operating procedures are followed. Further, hourly workers interact directly with customers in their own plant to solve quality problems where they arise.

Specialization Versus Work Teams

There is a corollary to this rule about decentralization that seems to fly in the face of conventional wisdom: When decision making is pushed downward, specialization can become counterproductive. To appreciate this, one only has to turn to the master of specialization himself, Adam Smith, writing in *The Wealth of Nations* (1776).[9] In his example of the pin factory, Smith recognized two factors determining the gains from specialization. Of these, the first and most widely remembered is that specialization is limited by the extent of the market. Thus in large plants there should be plenty of opportunities for workers to be devoted to one task, and the scale is such that their time in that task will be fully utilized. That is, they will not be idle.

The second limiting factor in Smith's analysis is setup time. Unfortunately, the importance of this side of his analysis is often neglected. Each time a worker switches tasks costs are involved in that the changeover can take time, and the worker is "unproductive" while making the transition to the next task. The implication of this is that the gains of specialization require a high degree of coordination in the workplace. But if decentralization is needed, that coordination becomes extremely difficult and costly. When a problem arises, continuous operation is promoted if it is confronted in a timely way. Waiting around for a specialist does not help to achieve that goal.

On the shop floor, the introduction of work teams facilitates decentralization. Within the team, workers are cross-trained (that is to say, not specialized). Teams are valuable precisely for the reasons we have indicated. Human capital is less likely to be idle, and decision making can be much more timely. When a coworker is absent close substitutes stand ready for work, and this substitution is made possible by cross-training. Consequently production lost because of the absence of any one worker is minimal. A firm's investment in this kind of organization can be substantial. Teams empowered to make decisions can take years to set up well. Training programs have to be established and evaluated. Workers have to certified as competent in a number of skills.

All this takes money. But note that the return on this investment is tied to the need for decentralization, and therefore, a firm's investment in continuous casting or ladle metallurgy and the accompanying computer hardware and software enhances the value of its investment in its employees. Similarly, however, a firm's investment in its employees and in the way their work is organized can make its investment in the caster or the vacuum degasser that much more valuable, as more metal is pushed through and wider markets are served. *Investment in these technologies and investment in people and in work teams are complementary.*

It is worth noting that the decision to invest in the development of empowered teams at steel mills was not typically a conscious acknowledgment of their complementarity with the new technologies. Indeed, talking to the managers at mills where teams are in place and effective, the incentive for their action is traced back to a simple recognition that rules limiting the way a worker can spend his or her time historically resulted in egregious waste and idleness. Steel executives and plant managers wanted to eliminate this just because it was an obvious source of productivity gain and cost savings. They pursued that kind of efficiency aggressively, just as they closed down redundant plants for the very same reason. Moreover, the investment in a better trained, more flexible workforce empowered to make decisions on the plant floor often came at tremendous cost—after long strikes or broken unions, for example. One does not make costly investments unless the expected return is also great. Whatever the motivation, however, the total effect of these decisions was to enhance the return that would be realized by investments in new steelmaking technology.

Combining Human Resource Practices

There are complementarities among various types of human resource practices that have been similar in importance to the complementarities we have described between labor and capital. When a company relies on work teams, for example, it becomes much more difficult to monitor the effectiveness of individual employees. It has long been recognized that work teams have the effect of transferring at least part of the responsibility for the monitoring of any given employee to peers in that worker's team. If a worker lags, it is to the advantage of other team members to correct the problem. Pay-for-performance systems that reward groups (e.g., teams or larger groups within a plant) enhance the incentive for workers to monitor one another and penalize peer behavior that would diminish the effectiveness of the team. Here is an instance, therefore, where a decision to implement a pay-for-performance system could be expected to increase the productivity of teams—two distinct investments in human resource practices that are mutually supportive.

The evidence that complementarities in human resource practices enhance productivity is both qualitative and quantitative. In 1989, the Brookings Institution organized a conference to examine the research findings to date on whether any tangible link had been established between labor productivity and particular pay systems or employee participation in decision making. Alan Blinder was asked to be the organizer, and a number of papers were commissioned to review the state of knowledge in various related areas. In summarizing the papers from this conference Blinder pointed to one wholly unanticipated thread of reasoning that could be traced as a common element in the research presented: arrangements that involve employee participation seem to make pay schemes like profit sharing more effective as incentive mechanisms.[10] Productivity gains are magnified by the interaction between these practices.

More direct statistical evidence on the importance of such complementarities in the steel industry comes from the recent work of Ichniowski, Shaw, and Prennushi.[11] These researchers, working as part of the larger CMU-Pitt steel project, gathered data from twenty-six steel plants. On the basis of site visits and questionnaires they compiled production data on one very specific steel production line to ensure comparability across their sample. By "holding technology constant" in this way, the study team was able to focus on human resource practices as the cause of productivity differences. The team concluded that plants that had implemented a full system of innovative human resource management policies were more productive than those that did not. Thus the complementarity and self-reinforcing effects of multiple programs all supporting behaviors in the desired direction are much more likely to result in the improvements desired than the "program of the month" approach, which is not supported by other changes in the work environment. Taken as a system, the totality of human resource management practices in a plant matter quite a lot. Taken one at a time, their value is diminished substantially.

Coordinating changes in the work environment is not easy. Change requires people to do their jobs differently, therefore it will be resisted. This is particularly true in the case of the lowest level of management, the front line supervisor. As decision making cascades downward and employee participation is stressed, the jobs of supervisors (or foremen, as they were called in the past) are at risk. When operating teams or crews are given greater responsibility for their own area, supervisors are no longer needed to play their traditional role, which was to bark orders at the hourly workers. Supervisors are necessary only to the extent they can facilitate the work of the crew. Order giving is out, and coaching, facilitation, training, and team building are in. Needless to say, this level of management is seriously threatened by the types of changes that are going on in the workplace today, and this phenomenon is not isolated to the steel industry by any means.

To succeed, one has to recognize the basis for this resistance and counter it. At USS, for example, the training available to supervisors has increased in areas such as team building, conflict resolution, motivation, and communication. Moreover teamwork is being recognized as one of the criteria in promotion and merit pay decisions. In effect, the company is trying to help these individuals grow into their new role. If that does not work, there is no real alternative to reassigning the supervisors or letting them go.[12]

Culture and Exceptional Performance

The complementary investments that we have described combine to produce outcomes that help to define the culture of a workplace. Firms that seize opportunities to capture new markets by using technology to gain a strategic advantage enhance their chances of success, and their profitability, by exploiting opportunities to tie changes in the organization of work to their market strategy. The vision and values of top management set the goals and the strategy, and once this is in place it is a small step to identify the capital investment necessary to gain access to markets and maintain them through cost competitiveness. But decentralization of decision making, reducing the complexity of a firm's organization, and building an environment where the incentives at all levels support the strategic goals of a firm are critical elements of success in the long term, precisely because they make investments in new technology that much more valuable.

A company's culture establishes the norms of behavior that determine how individuals and teams carry out their daily responsibilities. It is something that must be developed and nurtured by the leadership of a company, and in which one must invest scarce resources, time, and money.

If decentralization is important to a firm, then the culture that is established becomes fundamental to its success. The reason for this is that every decision maker has to be counted on to take actions that will enhance the value of the firm. Even if workers have bought into the firm's goals, they must know that there are guiding principles by which their own success or failure will be judged. If the uncertainty about that is too great, decision making will be inhibited; worse, the decisions taken may be counter to the best interests of the company. Trust in one's superiors and subordinates is the bedrock upon which all else is built.

We saw this in the successful minimills. The high value that top executives place on trust creates a company culture that stresses responsibility, accountability, and continuous improvement. With this culture driving the decision making process, these minimills have been transformed into high-performance organizations featuring significant decentralization. They have few organizational layers, little bureaucracy, and empowered decision makers. In these organizations, "good" decision making means fast decisions based on thorough discussion, but not necessarily time-consuming consensus decision making.

In explaining how this works, one of the supervisors in Oregon Steel's Portland melt shop said, "Expectations are well understood and communicated constantly; there is nothing ambiguous about what is expected of me in my job. We are thinly staffed and don't have time to waste. We discuss things all the time with our boss. If I need help he is always available. People don't feel left out of the decision-making process; we are always consulted and have the opportunity to make input into anything that affects us."[13]

Trust is built by a number of self-reinforcing policies and practices. These include openness, ongoing communication, involvement in the decision-making process, clear expectations, reward for performance, and control over many factors that influence that performance. Workers feel that their input is important, that they are listened to, and because of the openness and the high level of communication, they know what is expected of them and how their performance affects the goals and objectives of the company.

In discussing how "trust" develops, Rodney Mott, previously Vice President and General Manager of Nucor's Hickman plant and now in a similar position for the South Carolina strart-up, explained,

> There is no single, simple approach to building trust. Selecting employees who are capable of trusting others is necessary, and then open constant communication is essential. This helps to establish a working environment in which people understand what is expected of them, it helps to make people secure in their job and it gives them the confidence to make suggestions to improve things. I communicate by walking through the mill daily to talk to the operators; I meet with my managers at least weekly, and we have spontaneous discussions as the need arises; I write a weekly newsletter to keep everybody up-to-date, and to tell them what needs to be improved; I have dinner with each crew twice a year, and we have an annual summer picnic and a Christmas party. In addition, team decision making is the norm. The composition of the team will vary depending upon the issues under consideration, but input of all affected is encouraged and listened to. We often see people on the job for only two months actively contributing to problem solving and making improvements. We encourage this—nobody's afraid to make a mistake, and this attitude pays off for everybody.[14]

Trusting employees to do a good job, if given the opportunity and support to do so, is an important influence on the way these companies are organized. Decentralization and empowerment over the workplace flow from high levels of trust. With fewer organizational layers and little bureaucracy, information flows faster up, down, and across the organization, leading to faster and better-informed decisions. As employees see that their input is valued and that they are trusted to take control of their own workplace, they are able to see a clearer connection between their own actions and the results of those actions. Because employees have control over their own workplace, accountability for performance is easy to establish and linking rewards to performance reinforces the concept that what an employee does directly affects not only the company, but the employee as well. The concepts of ownership over the job and being rewarded for the way that job is carried out are important elements in establishing a culture in which continuous improvement is embedded into employees' approach to their jobs. These principles translate into job security for workers and help tie their long-term goals to the long-term goals of the company.

Communication is Key

Ongoing, open communication is a cornerstone of building trust and of ensuring that management's goals and objectives are well understood and embraced by all employees. As elaborated upon by James Todd, "In order to help build trust and commitment throughout the company, we have always been open with our employees. Even before we went public we would share our financial information with them. We have always had a policy of meeting with our employees on a regular basis to discuss anything they wanted."[15]

Being available to talk to employees is another component of Nucor's approach to business. John Correnti, President of Nucor, underscored its importance; "When I was a plant manager, the most important thing I did was to walk through the plant for an hour and a half or two and a half hours every day. People skills are the key to success. I didn't wear a coat and tie. I went to their place of work to listen to their questions and concerns and I still get calls here at headquarters."[16]

Communication is fostered in many different ways at Nucor. "Town meeting" dinners, similar to Tom Graham's "prayer meetings," are regularly scheduled between the plant manager and small groups of workers to provide an opportunity to directly discuss scheduling, equipment, organization, and production problems. Hourly employees are also included in customer visits to facilitate a better understanding of the client's needs, and participants are paid their base hourly rate for these trips. General managers hold annual dinners with every employee in groups of twenty-five to one hundred at a time to give employees the opportunity to discuss issues of importance.[17]

Communication about the worth of an individual to the organization takes many forms. Nucor, like Oregon Steel, makes every effort to take an egalitarian approach to the way in which people are treated across the organization. For example at Nucor, senior executives do not receive company cars, corporate jets, executive dining rooms, or executive parking spaces. In the plants all employees wear the same colored protective clothing and hard hats, all employees fly coach class, frequent flyer awards are redeemed for company travel, there are no assigned parking places, and all employees are listed on the cover of the annual report. All employees have the same holidays, vacation schedules, and insurance programs. In addition, a profit-sharing plan is maintained for all employees except officers.[18]

Access to information, assuming the information is accurate and well explained, supports a high-trust environment because employees know what is going on and they know that management is willing to share everything with them. Furthermore, an open communications process provides ample opportunity to reinforce the company's values and its goals and objectives, which in turn supports the culture. This gives management the opportunity to stress over and over again the goals of profitability, quality, accountability, reward for performance, and the relevance of continued improvement in these objectives. Thus the communication process facilitates the ongoing transmission of key values and objectives.

This works well as long as those communicating live up to what they espouse. In these minimills there was strong agreement that top management lived its philosophy and values—managers "walked the talk," as evidenced through the policies they implemented and their interaction with employees.

Summing Up

The ideas discussed in this chapter apply equally well to integrated steel producers and to minimills. In particular, the lessons drawn from restructuring in the steel industry indicate that success is tied in important ways to the ability of a firm to invest in technology and human resources so as to take maximum advantage of the complementarities that exist between them. In order to do this effectively, leadership is required to establish goals and build an organization and a culture that promote a firm's strategic interests in light of those goals.

Building the necessary culture requires the careful interlocking of a number of policies that motivate employees at all levels of the company to strive for improvement. Decision making has to be decentralized. To take full advantage of worker motivation and learning, flexibility in the workforce must be stressed. Teamwork is often an important part of this. Within these work groups, cross-training with respect to skills means that production workers can and do perform normal maintenance tasks, for example, and do not stand idle. Added to this gain is the fact that workers within groups are mutually dependent, and because income is tied to productivity and quality, there is strong personal incentive to cooperate and to promote the cooperation of others.

The value of investments made by firms in order to establish these practices in their plants is enhanced by modern steelmaking technology, which demands flexibility and promotes efficiency. In turn, the value of that technology to the firm is enhanced by these practices and the work culture they help to define. In the next two chapters, we shall find that the validity of this argument is not limited to firms in the United States. Our competitors abroad have learned this lesson too.

7

Regaining Competitiveness in the United Kingdom

Faced with severe excess supply conditions, eroding prices, and large losses in the 1980s the steel industry in the European Community restructured significantly, much as did their competitors in the United States.[1] In both places, when the integrated producers closed plants and strove for more efficient operation in their remaining plants and when minimills staked out new markets, the pattern in their actions was very clear. The most successful firms sharpened their focus on product lines, and used that as a guide for strategic investment. In the process, they reduced costs and enhanced product quality.

When an industry experiences common crisis on a global scale, there is valuable knowledge to be gained by comparing and contrasting the ways in which that challenge is met by firms in different countries. The nature of competition, the linkages between government and industry, and the relationship between a firm and its workers often have elements that are nation specific. As these conditions vary from place to place, the character of restructuring is certain to vary. However, just as the unique characteristics of a nation's experience are rich with opportunities to learn more about the forces that shape competitive advantage, important aspects of the competitive challenge and response cut across national boundaries. We can expand our understanding from the common ground too.

This chapter focuses on the ways in which British Steel, the major integrated producer in the United Kingdom, and Co-Steel Sheerness, the largest minimill producer in that country, have restructured to gain a competitive edge in their respective markets. The discussion examines the management practices used in restructuring these companies. Their experience confirms the broad applicability of the principles and insights we have gleaned from previous chapters.[2]

Underlying the restructuring that occurred at British Steel and at Co-Steel Sheerness is long-term commitment to organizational change. Both companies were guided in this by a strong market and product focus. In order to succeed, both companies also had to regain management's right to make resource allocation decisions on the plant floor. In the process each built an organization and culture that was committed to instilling the concept of continuous improvement within the workforce, and used innovative human resource practices to create a high-performing workplace.

Although there are some differences between American and British companies based on culture and unique company conditions, their experiences reinforce one another in terms of what is required to create a competitive company. Market focus, a linking of investments to those markets, and coordinated change in the work environment are necessary elements of competitiveness. Here and abroad, workplace practices were adopted that helped to bring about and support the decentralization of decision making. Firms strived for and secured greater flexibility in the workforce. Job classifications were reduced in number and cross-skilling was encouraged. Work became organized around teams, and financial incentives were put in place to promote performance on the job. There was significant investment in training, and merit was given a greater role in promotion decisions. Once again, we draw on company experience to examine the "how" and "why" of these changes.

Restructuring in the Integrated Sector

The British steel industry was nationalized in 1967 when fourteen companies were brought together. At that time, approximately 250,000 workers were employed and capacity was at the level of 20 million tons. Nationalization had been anticipated for some time and many of the companies involved had minimized investment and modernization in anticipation of the transfer of ownership. As a consequence British Steel was hardly a model of efficiency. The companies purchased by the government were overstaffed and the union was very strong, making it difficult for management to make changes in work practices and manning levels.[3]

At that time, government planners anticipated significant growth in steel demand, and they projected expansion of the industry to 35 to 40 million tons,[4] but this never did come to pass. Domestic market growth was slight and the energy crisis in the mid–1970s dampened demand further. By 1975 the company was experiencing large losses which continued for a decade, and the British government was compelled to subsidize the industry extensively.

Under the Conservative government of Margaret Thatcher, the decision was taken to return British Steel to private ownership, and in 1988 the government sold its shares to the general public. Along with privatization, the company's debt burden was forgiven—at substantial cost to British taxpayers and with substantial benefit to the new shareholders. But from that point onward, British Steel was cut off from government support, and along with a small number of other privately owned European steel firms, British Steel was placed in a competitive environment where failure could translate directly into bankruptcy.

It became obvious that rationalization of British Steel was inevitable early in the 1980s. The years that followed were politically tense, but the company rose to the challenge. In many ways the strategy taken was similar to that of the U.S. integrated producers. Inefficient plants were closed, manning levels were reduced in those that remained, and investment was concentrated in plants with a locational advantage—in this case, coastal plants. Within the plants, up-to-date operating practices were introduced. Unlike the large U.S. producers who narrowed product lines dramatically, British Steel chose to maintain a broad portfolio of steel products, but not all these products were produced in a single plant. The plants were reorganized and became specialized in particular product lines, for example, rod or flat-rolled, just as were those in U.S. Steel and other American integrated firms. British Steel had no intention of retreating from lower value-added markets inasmuch as it was the dominant producer in the U.K., satisfying almost 60 percent of total British steel consumption.[5]

With restructuring, employment fell from 186,000 in 1979 to 37,000 in 1994.[6] During that period productivity rose greatly and manhours per ton fell from 14 to 3.8.[7] Today this firm, once a moribund behemoth, is generally recognized as one of the most productive in the world.[8] Its competitive advantage stems from low labor costs (derived from low wage rates as well as good labor productivity), efficient manning practices, efficient hot metal practices, and its dominant position in its home market. Perhaps the single most important factor behind the transformation of the company into one of the world's best was management's ability to regain control of the workplace from the unions in 1980.

The Crisis and Labor Relations at British Steel

Up through the late 1970s, labor-management relations in the British steel industry were antagonistic and the unions essentially ran the workplace. Organized labor was reluctant to accept even the most minor change without long debate and sought to increase manning levels for any concession granted. This brought about excessive manning and rigid job classifications, with the result that the company was not competitive. Large losses and continuing excess capacity in the late 1970s made it obvious to management that significant changes in the organization of work had to be made if the company was to remain viable in the 1980s and beyond.[9]

In late 1979, the British Steel Board made a decision to reduce capacity from 21 to 15 million tons and to cut employment by a much greater percentage. Division managers were given targeted capacity cuts and employment levels, and were told to consult with union bargaining units at the plant level to work out the specifics. The magnitude of needed change was substantial. For example, the Strip Division, which produces the company's flat-rolled products, was told by the Board to reduce capacity by 50 percent. The two integrated plants in this division are located in close proximity to one another in South Wales, the Llanwern Works in Newport and the Port Talbot Works in Port Talbot. Together they employed nearly 22,000 workers, and were the lifeblood of the region's economy. The division's management discussed several options for achieving the cuts, and in January 1980 recommended to the Board a program of change that it called *Slimline*. Under that framework, more than 11,000 jobs would be eliminated over a period of only six months. Before it could be implemented, there was a national strike of all industrial grade steel unions in support of a national wage increase.[10]

The strike was settled in April 1980 with agreement to an 11 percent basic wage increase and a lump sum bonus to be negotiated locally that would be at least 4.5 percent. However, the agreement stipulated that all pay increases had to be financed through improved performance at the local level. This stipulation was coupled with the knowledge that the company was going to downsize its operations significantly. And there was the implied threat that if a particular plant did not perform well it may be closed.[11]

Coupled with the resolve of the Thatcher government to strip power from national unions, these realities strengthened management's position and ultimately returned control of the workplace to management. The wording of the agreement at the works in South Wales is indicative of the negotiating atmosphere in which this outcome was determined:

> Having regard to the grave business difficulties, the heavy burden of fixed costs, the Corporation's decisions not to continue with unsustainable overcapacity of plant for Strip Mill products and the over-riding need to become and remain internationally competitive given the alternative of closure, the signatory unions on behalf of their members commit themselves to full cooperation in the operation of the reduced plant configuration, to undertake all work necessary to sustain optimum plant performance and the implementation of competitive manpower practices and levels.[12]

Management Exercises Authority

British Steel's management made the most of the opportunity presented. They won and exercised the right to move workers between departments, to discontinue unnecessary operations, to contract out, to reduce the number of workers, and to introduce automation where feasible. The union could negotiate where reductions occurred but not the level of the cuts. In addition, significant changes in work practices were specified. Some examples of these changes reveal the depth of the problems involved: lifting a tap hole drill beam (previously two fitters and two riggers, now two fitters); changing an electrical motor (previously two fitters and one electrician, now one electrician); clay gun maintenance (previously two fitters, two fabricators, two riggers, now two blast furnace crew members); road transport (previously the Port Talbot Works and Llanwern Works were divided into two union zones and drivers had to be changed to cross from one zone to the next, now there is complete mobility across the operations in South Wales regardless of union affiliation.).[13]

The importance of these changes to management and to the company's ability to remain competitive was underscored by E. W. Denham, Director of Corporate Planning, who said, "We would close a plant down if we lost the right to manage it."[14] In reality, the situation at their mills was such that the company had lost that right, as these examples of work practices would indicate. Regaining the right to manage was a necessary condition for survival.

This fact, too, cuts across national boundaries. Tom Graham was vigilant about never giving up the right to manage. Tom Boklund at Oregon Steel and other minimill executives were emphatic about it, and instilled that prerogative throughout the management ranks in their companies.

It is important to realize, however, that firms can be internationally competitive in a union environment. This is certainly true among integrated firms, but examples are also found in highly competitive minimills. Indeed, Boklund's Oregon Steel is a case in point. The union at Oregon Steel's Portland plant was broken only after a long and bitter strike, yet this company very recently acquired a unionized plant as part of a strategic investment program to secure new markets. The firm viewed the union plant as critical to its continued success, and went into the venture only after it was convinced that it could secure cooperation on critical issues. The point is that the company and the union recognize that there is no real alternative to working together toward building a prosperous steelmaking operation.

The issue at British Steel or at Oregon Steel's Portland plant before the union was broken was simply that management had lost the right to make critical resource allocation decisions on the shop floor. In a highly competitive environment, the right to manage is the right to respond to challenges by initiating change. When that is lost a firm's vitality goes too, whether it is an integrated operation or a minimill.

Realizing this can be a critical step toward competitiveness. In restructuring, the first source of cost savings is often to identify and eliminate excess capacity. When plants are underutilized they have to be reorganized or closed to eliminate waste. Similarly, idle or underutilized people are a certain sign of wasted human capital. Competitiveness requires that managers have the authority to eliminate such waste when it exists, and to prevent its existence in the first place, if that is possible. That may mean laying off redundant workers, but just as critically, it may mean using the workers who remain more flexibly and more fully.

Customer Orientation

A second key element of the strategy taken by British Steel was a new found focus on customers and markets. Managers began to realize that they had to meet customer requirements and expectations for quality, delivery, and service. This occurred at about the same time that executives of U.S. integrated companies were forced to the same understanding. With the advent of increased competition at home and abroad, these companies could no longer expect customers to take whatever they shipped, whenever they chose to ship it.[15]

To emphasize its customer orientation, British Steel reorganized along product lines. There are executive directors at the Board level for long products, flat products, and commercial/distribution activities. Each of its four main integrated plants now has a clear product focus. Two of these produce long products, with additional processing at three other sites. The other two integrated plants produce flat products, along with smaller sites that have special coating capabilities. Separate facilities at other plant locations take semifinished steels from the integrated mills and use these materials to produce a wide variety of final market products such as tubes and pipe, tin plate, narrow strip, and seamless pipe. Additionally, the company owns Tuscaloosa Steel Corporation in the United States, a minimill producing plate in coil and cut form, and it holds equity positions in joint venture companies producing stainless steel, engineering steels, wire rod, and wire products. Recently, British Steel also entered into a joint venture called Trico Steel with LTV and Sumitomo Metal Industries to produce strip products in the United States.[16]

Having plants specialize by products makes it easier to focus on specific markets and particular customers. And this kind of specialization increases the likelihood that capital investments will be driven by customer requirements, an advantage inherent in minimill plants because of their smaller size and relatively narrow product range. Separating the finishing facilities from the steelmaking operations reinforces this focus and increases the level of accountability between units. This, too, enhances performance.

The company's market focus was sharpened by more stringent customer requirements in the automotive sector, one of its principal markets, and this spilled over to other products and markets. Increased competition and continued excess capacity in most of its markets has also provided a strong incentive to work more closely with its customers. For example, it is making a concerted effort to maintain its market position in the container business through the development of an easy-to-open steel can, and it is also aggressively trying to market steel frames for housing.[17]

Innovation in Human Resource Practices

After the strike was settled, British Steel pursued a strategy of adopting innovative human resource management practices. Management wanted to create an environment where workers would be highly motivated and manning practices were efficient. Some of the steps taken included reducing the number of job classifications, cross-training workers, and introducing more of a teamwork environment. The bonus system was also refined to include elements based on productivity, quality, delivery, and cost. This extended what had been a tonnage-based system to include factors essential to the company's competitiveness. The bonus system is based on the concept of pay for performance. It has to be financed by productivity improvements or savings achieved through cost reductions.[18]

Managers at British Steel see the implementation of these human resource practices as essential to achieving ongoing productivity advances and cost reduction. The company has made significant progress in reducing the number of job classifications, moving toward multiskilled maintenance workers, and using teams to carry out specific tasks or solve problems. The emphasis is on the development of cross-trained workers paid on the basis of their competencies. Eventually the plan is to go to self-directed work teams with the team leader being a working member. This advance is currently being carried out at several sites.[19]

The self-directed concept is well structured, and within it clear bounds are placed on decision making authority. Production goals are established, and in meeting those goals customer specifications have to be maintained to close tolerance. Success requires managers who are capable of working in this environment, a cooperative union, and careful attention to designing teams that have the right set of skills to carry out the job. Management has sold the concept on the basis of two factors: Customers will be better served, and job security will be strengthened.[20]

The foundation of the company's human resource policies is the development of competency standards for each job. Over a five-year period this was accomplished for every job in the company. The task involved was truly enormous and required a major investment in terms of time and talent. To make it pay off, an extensive training program was developed. The combination provided workers with clear knowledge of the skills required for advancement and the support they needed to acquire those skills. It was a dramatic statement by the company that competency mattered, and it also demonstrated that the company was determined to raise productivity by investing in its employees.[21]

Annual training plans have been developed as part of each plant's annual operating plan. Training officers within the plants facilitate the plan's development and its implementation. Management employees are assessed annually, and a training plan for each individual is developed as part of that process. Job training manuals have been produced for all production and service personnel and are updated by departmentally based unit trainers. Competency standards are built into these manuals and operators are tested and accredited prior to moving through promotional stages. The company has extensive entry-level training and apprenticeship programs, and numerous off-the-job course offerings to satisfy core skill and knowledge requirements. These include management and supervisory courses in basic management, industrial relations, costing; engineering-related courses in hydraulics, welding, electronics; and training in operating equipment such as a fork lift or crane. In addition, the company encourages employees to take courses offered by other organizations in steelmaking, rolling, languages, chemistry, and so on. In the past year the company invested approximately $2,000 per employee in training, and in the aggregate this represents a significant part of its total wage bill.[22]

With the increase in teamwork, the growth in competence testing, and the use of management selection, seniority is playing a decreasing part in the day-to-day manning of the plant. Employees have to be properly trained and demonstrate competencies for their job at all times. Merit is a strong consideration for any opening, and management can veto the recommendation of the union for any position.[23]

This comprehensive approach to human resource management has helped to embed the concept of continuous improvement into the work culture. It is supported through benchmarking (Port Talbot uses its primary competitors Sollac, Thyssen, and Hoogovens), the bonus system, training/competencies, promotion by merit, an orientation to processes and customers, and the specific performance targets that are established at the plant level for costs, productivity, and quality levels (these are reviewed quarterly). Underlying all this is the ever present threat of downsizing and job loss—survival is still a strong motivator.

Focus on the Workplace

Unlike Tom Graham's market-driven strategy for downsizing the United States Steel Group of the USX Corporation, the strategy adopted by British Steel was more cost driven. The market decision was easy—British Steel elected to remain in all its markets. What it needed to do, however, was to bring its capacity in line with demand, the same challenge facing all the U.S. integrated producers. And it had to run its remaining facilities as efficiently as possible. The latter challenge was emphasized by Tom Graham for all the companies he managed. Once the closure decisions were made, the imperative shifted to becoming the lowest cost, most productive producer providing the quality demanded by customers in each market segment.

The keys to British Steel's success have been its ability to establish control of the workplace and the progressive human resource policies that evolved throughout the 1980s and early 1990s. The massive redundancies and plant closures that occurred at the outset of restructuring the company contributed to an atmosphere of crisis. This enabled management to bring the workforce in line with globally competitive standards, and provided the incentive necessary for ongoing productivity and quality improvements. The human resource policies developed by British Steel were self-reinforcing with many of the same characteristics as high-performance workplaces in the United States. All these actions complemented the company's focus on customers and product quality, and they enhanced the return that the company could realize from critical capital investments.

Similar to the strategies and practices used by Tom Graham to make USS and AK Steel highly productive, those employed by British Steel illustrate that competitiveness requires a comprehensive approach on the part of management to all aspects of the way a company is managed and its strategies are set. In addition, the British Steel experience demonstrates that the road to competitiveness can be lengthy. Since the strike in 1980, which marked the beginning of major change in the company, British Steel has remained committed to improvement for nearly fifteen years. That kind of effort can only be sustained if real leadership is exercised and a culture of competitiveness is ingrained. British Steel has helped to establish that culture by its intense commitment to improving the productivity of its workforce through skill enhancement and by its complementary human resource practices.

The Transformation of a Minimill

Co-Steel Sheerness is a subsidiary of Co-Steel, Inc., a company that owns minimills in Canada, the United States, and the U.K.[24] The plant at Sheerness has the capacity to produce one million tons of finished steel products per year. It shipped more than 800,000 tons in 1994 and employed about 650 workers. The plant's product line is comprised of concrete reinforcing bar and rod for construction, angles and structurals for fabrication and equipment manufacture, special quality bar and rod for the automotive industry, and forgings and wire drawings for various industrial and consumer products. Production is based on standard minimill technology: an electric arc furnace wedded to a billet caster. And like many of its American counterparts, Co-Steel's product line has broadened in recent years in order to serve new markets.[25]

The company was started in 1972. During the remainder of that decade it operated in the traditional industrial relations mode characteristic of British manufacturers. It had an autocratic management style with little involvement by the workers. Promotion was by seniority, and consultation/negotiation was required with four unions over a myriad of issues involving day-to-day operations.[26]

However, during the national steel strike in 1980 management was able to convince the workers that it was in their collective interest to remain working in order to continue to supply customers and to ensure future job security. But that labor climate did not last. The recession of the early 1980s reduced demand and caused large losses, forcing a permanent reduction in the workforce of about 20 percent.[27]

Recognizing the Need for Fundamental Change

The consequences of this recession sent a signal to management at Co-Steel that they needed to find new ways to enhance competitiveness, and that the change had to be broad based. Although they did not define it as such at the time, they embarked upon a process that they would later describe as cultural change—a process that would carry on for the next decade and produce a high-performing workplace.[28] Hugh Billot, Personnel Director, said, "We knew that we had to break down the barriers between hourly and salaried employees in order to increase productivity."[29] Although the specific steps taken were unique to Co-Steel, the objective was similar to that at Nucor or Oregon Steel; that is, they had to build a culture that bound the interest of the workers to the company's goals. Along the way, they made Co-Steel more productive and more profitable.

Benchmarking was an important first step in the process. The company compared itself to the best in the world to find out how other companies were achieving their superior performance. One of the conclusions reached was that these companies were investing in people and making enormous gains by upgrading the skills of their employees. To follow up on this, the company selected a "best-practice" plant and entered into an agreement to send 20 percent of the melt shop employees to that plant for two weeks. Co-Steel's objective was simple: They wanted to learn the other company's practices and philosophy and transfer that knowledge back to Sheerness. Groups of other workers followed. This helped not only to change work practices, but to establish a continuous improvement mentality in the workforce.[30]

Training, Cross-Training, and Work Teams

From the benchmarking exercise it was apparent that many successful companies had developed a workforce that was cross-trained, which made it possible for them to employ fewer workers to accomplish a given task. These companies provided a high level of training so that their employees could perform a variety of jobs, unconstrained by artificial boundaries. Co-Steel began negotiating with the unions to provide the operators with training in maintenance in order to free up the craft workers to receive cross-training. This program has now evolved to the point where the company has individuals throughout the plant who are trained in both electronics and mechanics. Because of the versatility of these workers, it has been possible to reduce staffing levels dramatically.[31]

As part of the program to cross-train people, the company was able after five years to eliminate promotion lines and seniority concepts. This increased flexibility in the use of labor and provided additional labor savings. In the melt shop, for instance, there is only one job classification. All employees are called steelworkers and the rate of pay for each worker is determined by that person's skill and performance. Promotion to supervisor is also based on merit, instead of seniority.[32]

Significant training was required to support the development of a flexible workforce. Like British Steel, standards were developed for performing every job in the company, and there has been a close reliance on competency testing. Progression from one job to another is based on skill attainment and additional training is encouraged—much of it provided at company expense.[33]

In 1988 a Performance Incentive Plan was developed. The objective of the plan was to focus employees on the need to improve company performance, especially plant yield. Base levels were set for every operating unit and improvements beyond that level were shared between the company and the workers. In order to determine how each individual or team should share in the gains, an annual appraisal program was introduced for all hourly and salaried employees. This was based on criteria deemed essential to good performance, such as safety, training, absence, attitude toward work, and job performance. At the same time as this incentive plan was introduced, employees were no longer required to punch a time clock. The company also has a profit-sharing plan, and 8 percent of the pretax profits are distributed to the employees. Thus the link between reward and performance is clear and strong.[34]

The management at Co-Steel has made a serious commitment to the development of work teams in order to improve decision making and achieve greater coordination across the plant. Each of the business units within the plant is run by an interdisciplinary team of managers. All supervisors and shift managers have been put onto a team, with the shift manager becoming the team leader, responsible for helping those on the team raise their skill level. The team determines who does what job in each shift and there is significant movement between jobs. In the steelmaking operation, for instance, each employee works a different job on a weekly basis. The individual operating the furnace from the control room one week will be on the floor in front of the furnace the next. Team members are paid according to their skills, which provides an additional incentive for employees to seek out training opportunities.[35]

To a Nonunion Shop

Most of the workplace practices we have described were implemented in a union environment. Nevertheless, Co-Steel's management viewed the union as a serious impediment to further progress. In its opinion the union had contested almost every issue and had kept management from working directly with the employees. New practices had to go through union representatives, which presented obstacles to competitiveness both in terms of needed skill enhancements and in efforts to motivate the workforce.[36]

The groundwork for changing this environment was laid over a decade in which a number of small but important steps were taken that helped to reduce the barriers between management and labor. While these steps were not designed to break the union, that was the ultimate outcome. One of the key events occurred in 1988 when the union position of lead hand or first operator was upgraded to a working salaried supervisory position. This required the workers involved to resign from the union. About the same time operators were invited to attend shift management meetings at the beginning of each shift. This got them more engaged in planning and organizing the work activity. Promotion by merit and the performance incentive program further eroded the power of the union. Key members of the joint union/management committee were taken abroad to visit other steel plants to meet with union and management representatives and to learn more about different work philosophies, reward systems, organizational structures, and the like. In addition to the knowledge gained, the trips offered an opportunity for the salaried and hourly members of the joint committee to socialize.[37]

In early 1992, after assessing the changes that had occurred over the previous years, Co-Steel's management saw and seized an opportunity to create a single-status (i.e., nonunion) company. They developed a strategy for forcing the change and laid out their objectives in a weekend of intensive meetings with the plant's union committee. Workers were given an opportunity to join the staff as salaried employees with a medical plan and improved benefits. Half signed on within ten days and all but one accepted the change by the company's deadline. Ultimately, the union was decertified.[38]

Corresponding Organizational Change and Attention to Markets

Important organizational changes reinforced Co-Steel's cultural transformation. There used to be nine levels between the managing director and the operators. There are now four. Profit centers were created for the furnaces, casting, rod mill and bar mill, and each of these areas is run as a business by a team of managers comprised of a mechanical engineer, an electrical engineer and a production person. The purposes of these cross-functional teams are to increase the level of communication between the engineers and operators and to engender a business orientation into the operation of the plant, so that profits, cost control, and marketing can be stressed. These individuals are responsible for working with customers and suppliers, and they arrange worker training. Maintenance functions were also centralized, which means that maintenance workers can go wherever they are needed in the plant.[39] In addition to these internal changes, increased competition forced the company to become more marketing oriented and to increase its customer focus. Rather than operating as an order-taker selling commodity products, it has tried and succeeded in differentiating itself and its products in the marketplace.

Co-Steel has been in a profit squeeze. On the cost side, when the U.K.'s energy companies were privatized, energy prices increased significantly. However, the price of reinforcing bars (rebars), one of the company's core products, has been constrained by competition from Eastern European steel companies and countries with growing steel industries, such as Taiwan. Co-Steel's strategic reaction was to withdraw at least partially from the rebar market and move into higher value-added steel products. Now over sixty percent of the steel produced by the company is for these markets.[40]

Co-Steel's market strategy is similar to that of the U.S. minimills, who have also diversified into higher valued, more sophisticated products. By expanding its product mix to include flats, angles, channels, and special quality rod and bar for more sophisticated applications and offering all these in a wider variety of sizes, it moves closer to meeting the special needs of particular customers. Because of this it has adopted a more proactive strategy of working with its customers in order to identify the product characteristics that will serve them best. At the same time, that kind of ser-

vice helps tie the customer to Co-Steel.

Some of its steel, for example, is used to make roofing bolts for the coal industry. In order to better understand the conditions under which these bolts are used, Co-Steel is working with Britain's National Coal Board to see if there are better ways to meet their needs. In this case and others like it, the objective is to better understand what the customers want, and to initiate joint product development programs with key customers. This is the cornerstone of its strategy for moving into niche markets with higher profit margins.[41]

To further strengthen its working relationships with customers, Co-Steel invites them into the plant to discuss process-related issues and to understand what properties are critical to the customer for different applications. In addition, operators are sent out to the customers' plants to gain a better appreciation of how the steel is being used.[42]

Like its U.S. counterparts, customer requirements are also driving some of its capital expenditures. Productivity and product quality have been enhanced by the introduction of a new ladle furnace. In the rod mill $1.5 million was committed to a new cooling table. This investment makes cooling more uniform and thereby ensures more uniform properties in the bars. So when a customer draws the bars into wire there will be fewer breaks and thus increased yield. As a result the customer's profits are enhanced and Co-Steel strengthens its customer base.[43]

The changes introduced by management at Co-Steel have been instrumental in creating an environment in which attention to productivity, quality, and customers are a part of the culture. At the same time they have brought about a flexible, cross-trained workforce with greater employee responsibility, and an organizational structure that is more streamlined. All these changes have helped to lower costs and improve efficiency. Their success can be measured by a consistent decline in injuries, a consistent increase in tons produced per employee per year (up by a factor of 2.5 since 1980), improved quality, and greater profitability.[44]

The key to the company's transformation has been the creation of a high-performance workplace. As we have seen in so many other examples, the success of this firm can be tied at least in part to the implementation of a number of reinforcing policies. Collectively, they have moved Co-Steel to the point where the quality of the workforce and the management of the workplace combine to give it a competitive edge.

Summing Up

There are differences in the restructuring programs undertaken by British and American firms which arise primarily from variations in the competitive conditions confronting them over the past decade and a half. However, the overall approaches taken are very similar to one another in important respects. In both countries, the firms focused on productivity and cost reduction, they became much more oriented toward serving their customers, and they created a high-performance workplace by instituting a set of interlocking human resource practices. The successes achieved are a result of the complementarities among these strategies and the fact that they were implemented as a whole so as to take full advantage of the investments made in each.

Like the U.S. producers, the British companies are striving to improve both their processes and their products. They are not content to stand still and are focused on gaining and sustaining a competitive edge. Competition has been the driver behind these changes, just as it was in the United States, and sound management practices have been the solution, just as they were in America.

8

Sustaining Competitiveness in Japan

Until very recently, the pressure for change in the Japanese steel industry stemmed primarily from the worldwide capacity glut. Steel prices were forced downward on international markets, and Japan's foreign competitors were becoming more efficient because of their restructuring. This meant that Japan's advantage in export markets was being eroded. Japanese steelmakers, like all others, had to respond by seeking ways to reduce costs. But the Japanese producers faced an added burden brought on by the strength of the yen, which made their steel more costly when priced in terms of dollars, pounds sterling, or francs. As a result, cost cutting in Japan became ingrained. Now, however, the powerful integrated steel firms in Japan are being forced to react to two new threats—this time in their domestic market: incursions by firms from the rapidly developing nations of Asia and by progressive Japanese minimills.

Japanese efforts to control costs and the reaction of Japanese firms to competitive threats take place in an institutional environment that is radically different from that in the United States. In this chapter, we draw primarily on the experience of Nippon Steel Corporation,[1] the largest steel company in the world, to understand the nature of restructuring in Japan's integrated sector, and upon the experience of Tokyo Steel Manufacturing Company (Tokyo Steel),[2] the largest minimill in Japan, to learn more about this side of competition in the Japanese steel market.

As in Chapter 7, we focus here on management practices employed in the restructuring process. Change has not been as urgent in Japan as in the West because the financial pressures have been less severe. There are also two other features of restructuring in Japan that have helped to make their experience unique. First, the behavior of major steel producers is governed by Japan's widely known social compact, which entails lifetime employment in the larger companies. Second, Japanese firms have a long and rich history of innovative human resource practices. Many of the workplace characteristics that firms in other countries strive for today have been a normal part of the work environment in Japan for some time. As a consequence of these unique features, Japanese restructuring embraces only part of the comprehensive change that we have found elsewhere.[3]

Integrated steel companies in Japan have increased their efficiency through a steady and consistent focus on cost reduction—a focus that is not at all new to these firms. For them, the more dramatic side of change has been organizational. Japanese integrated steelmakers are reducing the size of their corporate bureaucracy, cutting out management layers, combining jobs and speeding up decision making by forcing it down further in the organization. Moreover, to provide incentives for employees to take on more responsibility and to become more innovative, Japanese producers are paying more attention to performance-driven bonuses and to promotion by merit. These things are new and challenging for the Japanese, but the pace of change is slower than it was in the United States or Europe.

The same competitive pressures confront Tokyo Steel, the Japanese equivalent of Nucor. But because of Tokyo Steel's streamlined organization, its emphasis, like Nucor's, has been on innovation. This firm is moving into high value-added products by attacking markets of the integrated producers, and it is linking rewards to performance. Unlike Nucor, however, Tokyo Steel has encountered stiff competition from the integrated firms, who have resisted its incursions into their markets and refused to pull back quickly as did the large U.S. producers in the mid–1980s.

The Context for Japanese Restructuring

From the mid–1970s through the 1980s, Japanese steel companies reduced employment by about 40 percent or 152,000 workers,[4] with many of these workers finding jobs in affiliated companies. The downsizing, however, was carried out in a more orderly manner than that in the U.S. and Europe where employment losses for a comparable period were 60 percent (315,000 workers) and 56 percent (494,000 workers) respectively.[5]

Figure 8.1 shows Japanese and U.S. steel production over this period. The comparison is revealing of the differences in the magnitude and timing of restructuring that was required. The pattern in Japan is clearly cyclical, with no strong upward or downward bias in the long-term trend. The volatility of U.S. production and its downward trend are evident in comparison to the Japanese experience. Moreover, the devastating collapse of U.S. production in 1982 marks an unmistakable turning point, whereas there is little in the Japanese data to suggest crisis of a similar magnitude.

Just as the Japanese government played a partnership role in helping to secure initial success for the steel industry, so also it stepped forward to guide and facilitate the gradual restructuring of the industry. Legislation

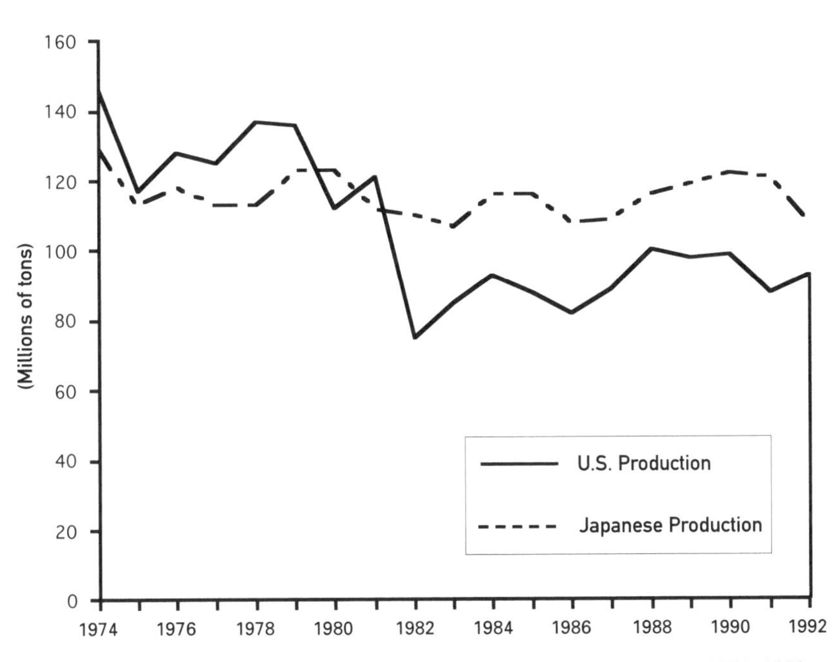

Figure 8.1. Raw steel production in the United States and Japan, 1974–1992. *Source:* American Iron and Steel Institute. *Annual Statistical Report* (Washington, D.C.: American Iron and Steel Institute, various years).

was passed in the late 1970s and early 1980s that empowered the Ministry of International Trade and Investment (MITI) to coordinate an industry-wide process for planning the reduction in capacity, which was the industry's first step in cost containment. Efforts focused on assisting the integrated producers in diversifying their holdings to nonsteel enterprises as a way of opening up job opportunities for redundant steelworkers. Low interest rate loans and loan guarantees were made available to help firms establish new lines of business. Assistance to fund allowances for early retirement, and subsidies for worker retraining and the costs of temporary layoffs were provided. In addition, the government administered an employment adjustment subsidy program, funded by the private sector through a payroll tax, which enabled companies in industries experiencing declining sales to receive up to two-thirds of the wage bill for "seconded" or transferred employees. In effect, the policy transfers income from growing sectors of the economy to firms that are being forced to reduce employment. This has dramatically decreased the financial commitment associated with restructuring in the steel industry.[6]

In terms of its influence on capacity and investment decisions, the role played by MITI in the 1980s and 1990s toward the steel industry has been far less proactive than it was in previous decades. In part, this simply reflects the fact that the industry was well established by that time, while other industries had become priorities for Japanese industrial policy. However, MITI still sets voluntary production guidelines for the industry and provides a forum to bring together industry leaders, government officials, bankers, and other parties interested in the industry through its councils and subcommittees.

The importance of these meetings should not be underestimated. They enable MITI to exercise administrative guidance in a process that is difficult for outsiders to understand. In this way, MITI has been very effective in helping the participants gain consensus about what is in their collective best interests, where the issues on the table may involve pricing, output, diversification, trade, joint research, the environment, or rationalization.

Although MITI cannot order companies to take specific actions, the ongoing nature of these interactions and the wide variety of issues discussed over time make it difficult for participants to go against the group consensus. Each party realizes that it will need the support of the others at some future time, and therefore there is a give and take among those involved that is unusual by Western standards. This consensual decision making between industry, government, and other stakeholders gives an edge to the Japanese industry compared to its U.S. and European counterparts, in terms of the formulation of policies that affect the industry as a whole.[7]

During the 1980s and early 1990s, the Japanese steel industry focused on achieving greater operating efficiencies. The initiatives involved consolidating blast furnace operations, going to almost 100 percent continuous casting, upgrading finishing operations, and making improvements in all facets of the manufacturing process.[8] There were productivity gains as a result, and the need for many blue collar employees was eliminated. But in one important respect, the restructuring options available to Japan's integrated steel producers are more limited than those in the United States or Europe. The social compact between large steel firms and their workforce that guarantees lifetime employment for full-time workers means that plant closures and layoffs are not the normal industry response to competitive pressure.

In order to absorb redundant labor and to diversify into high-growth sectors, most of the integrated Japanese steel companies started new businesses in the 1980s. Nippon Steel, for instance, formulated a strategy to diversify into advanced materials, chemicals, electronics, communication systems, and engineering. [9] Other companies branched out into semiconductors, computers, and real estate development. Measured by the profitability of the new ventures, this diversification has not been an economic success in many cases. But the new lines of activity did provide job opportunities that were used to absorb some of the excess steel employees.[10]

By 1994, the major steel companies had transferred a large percentage of their steel employees to subsidiaries or other businesses. As one case in point, Nippon Steel shifted about 15,000 workers in this way, about half of its steel workforce. While producing some savings, the company was still responsible for making up the difference for each employee between his or her salary before and after the transfer, which normally would have included an employment adjustment subsidy. This amounted to annual expenditures exceeding hundreds of million's of dollars.[11]

Unprecedented Change in Japan's Integrated Sector

In the early 1990s, the Japanese industry was awakened to the fact that business as usual would no longer work, a wake-up call that was received a decade earlier by U.S. and European steel executives. The rising value of the yen, a recession at home, increased competition from domestic minimills, and rising imports combined to produce negligible profits in fiscal 1993 and large losses in fiscal 1994 and 1995.[12] The practice of reducing employment by transferring employees to other organizations was no longer as viable an option as it had been in the past. Many of the affiliated companies and supplier organizations were not growing and were confronted by the same competitive pressures that were being faced by the steel industry.

Aggravating the problem was the fact that the industry now had significant excess capacity, estimated to be about 20 million tons,[13] and foreign steel could be sold in Japan at prices significantly below that of Japanese steel. The seriousness of the situation was revealed clearly in late 1994, when Mitsubishi Motors took the unprecedented step of entering into a long-term agreement with Korea's Pohang Iron and Steel Company to buy sheet steel for car production. The contract called for prices estimated to be 15 to 20 percent below comparable domestic steel.[14] This agreement turned up the pressure on Japanese producers to reduce costs, and highlighted the devastating consequences of the yen's appreciation on world currency markets. Even the tight-knit working relationships among Japanese companies, organized within a group of companies having common interests and known as a *keiretsu,* is not impervious to global competitive pressures.

The large Japanese steel companies are responding to this competitive pressure on a greater scale and in a more strategic manner than in the past. The integrated producers announced three-year plans in 1994 to reduce the size of their workforce still further and to become yet more efficient. The emphasis of these strategies embodies certain aspects of reengineering, namely eliminating overhead costs through organizational changes. A Nippon Steel executive explained that even though the company's steel operations were competitive with the best in the world, its "multilayered" overhead costs had undermined its profit potential. He emphasized the importance of decentralization and delegation. What is needed, he said, "is speediness in decision making. We are aiming to realize a small head office."[15]

Job Security in Restructuring

Nippon Steel's plan is designed to strengthen its international competitiveness by securing a reduction in annual production costs of 300 billion yen. In the process, the workforce will be scaled back by another 7,000 jobs in steel operations. Outsourcing costs will be reduced as one step toward achieving new operating efficiencies. Management systems are also being redesigned, with the objective of streamlining operations.[16]

Still, the company does not intend to lay off employees, and will continue to make every effort to find them jobs in subsidiaries, affiliates, or other companies. Given the fact that reductions have been in progress for several years, however, there is not much opportunity to place significant numbers in subsidiaries or affiliates, and therefore greater effort is being made to find them jobs in other companies. Nippon will guarantee the workers their present salary for three to four years and will pick up the difference between the current salary and that of a replacement job for that period.

While the company is morally committed to the no layoff policy, it has modified that position somewhat. In order to give workers an incentive to quit the company, Nippon has offered to pay some workers a separation bonus equal to the present value of what they would have earned had they continued working there until they retired at the age of sixty. Thus, economic pressures are forcing companies to find ways around the ironclad lifetime employment commitment without destroying the spirit of that commitment.[17]

Nippon Steel's extraordinary effort to make good on its promise of job security goes a long way toward explaining the trust that has been established between the firm and its employees. The union has been thoroughly briefed on all the reductions in force, and lends it support primarily because the company has agreed to ensure that other jobs would be found for redundant workers and salary levels would be maintained.[18]

These are trying times in Japan, and it is worth remembering that the social compact that binds firms to workers and workers to firms is a relatively new phenomenon that grew out of turbulent labor-management relations in the late 1940s and 1950s. In the period of sustained growth that followed, the compact went unchallenged. It is under a good deal of strain today, however, and will survive only as long as companies and their stockholders are willing to take less of a return in a down economy.

Reorganization and Decentralization

In the years ahead, the Japanese workplace will become more decentralized. Nippon Steel is reducing the number of management layers by one, and while this is an extremely modest goal by American or European standards, a large number of management jobs are being eliminated as departments and divisions are combined. Throughout the entire organization 156 divisions have been combined into 112, a 28 percent reduction. The purposes of these organizational changes are to produce cost savings, speed up decision making, and improve market focus as sales and technical divisions are combined into product groups.[19]

Decentralization is part and parcel of the reorganization, and decision making is being pushed downward. Nippon's steel plant management, for example, can now make decisions for higher levels of capital expenditures without approval from above. And efforts are being made to pinpoint precisely where in the organization decisions should most appropriately be made.[20] While all these changes are designed to improve Nippon's competitive position, the pace of change is slow and its magnitude at this point is small relative to that of steel companies in the United States or the U.K.

Teamwork and participation were always a part of the Japanese work culture, and the consensus decision making style that evolved was very time consuming and may have dulled individual initiative and creativity. The organizational changes that have promoted decentralization may help to counteract this and speed up the decision making process. The potential implications of this trend are very broad. In particular, Nippon's top management knows that as decision making is decentralized, greater attention must be given to reward for performance. Individuals cannot be asked to accept and exercise new responsibilities unless reward is linked to the result of their actions, especially when careers are on the line.

Compensation and Promotion

At present, Nippon's promotion and reward policies incorporate a mixture of seniority and performance as the criteria for wage increases, bonus levels, and promotion to higher job classifications. Traditionally, however, seniority has been the major factor in wage and promotion decisions, and it still plays a predominant role. The explanation offered here is that older workers deserve greater rewards because they have accumulated more knowledge by being on the job longer. Moreover, there is an equity side to the argument: In the Japanese view, older workers have greater economic need. Today, however, with technology changing faster and becoming more important in the workplace, time-on-the-job and competence are not as clearly related to one another as they were in the past, and change is in the wind.

The Japanese system for promotion is not very flexible. Advancement for blue collar, white collar, and management employees requires a minimum number of years within a given job classification before the employee can be considered for promotion. At Sumitomo Metal Industries, for example, an individual hired right out of college would have to put in a minimum of fifteen years in various lower level positions before becoming eligible for promotion to the level of plant superintendent;[21] at Nippon, management employees have to be in their first position for about eight years before they are eligible for promotion, and blue collar employees would have to put in at least twenty-two years before being considered for promotion to the position of group leader.[22] Merit enters the promotion decision only after the criterion of time-on-the-job is satisfied. Then an individual's merit rating can matter in the selection process, where merit is determined by the person's breadth of knowledge about the job and the work group as well as adjacent work groups, and the ability of that person to contribute as a member of a team.[23]

Semiannual bonuses are a very important feature of the Japanese pay system and can represent 30 to 40 percent of an individual's total income in a typical year. In practice, the bonuses are not closely tied to company performance, and they are only modestly tied to individual performance. For instance, between 1992 and 1994 when Nippon went from being extremely profitable to incurring large losses, the bonuses were cut only about 10 percent. In any given year the bonuses parceled out to workers are usually clustered around the companywide average in a tight range. But a small number may increase or decrease significantly from the average, say by 30 percent, based on assigned merit ratings.[24]

Promotion and compensation systems need to support organizational change. Good employees have to be open to change and willing to learn new ways of doing things, and this is particularly important in organizations such as the one Nippon is trying to create. As positions are eliminated or combined at Nippon, the content of many jobs is changing, and at the same time individuals are being asked to assume more decision-making responsibility.[25] Inevitably, as this happens the knowledge base required for good decision making is broadened. Those most able to adapt and perform well in the new environment have to be recognized and rewarded. Changes in traditional pay and promotion practices can therefore be expected to evolve toward more performance-based systems. However, change will be slow. The strong Japanese cultural beliefs in equity and fairness translate into expectations that all members of a team or group should share equally in the rewards, and this is a powerful restraining force.

The Outlook for Offshore Investment

Offshore investments in manufacturing facilities have been important for Japanese steel producers and this practice is likely to continue. Much of the incentive for this comes from the fact that Japanese automobile manufacturers and other Japanese steel users are setting up operations abroad. The auto makers and other firms have been attracted to foreign investment opportunities by phenomenal rates of growth in the emerging market economies of Southeast Asia and China. And the strength of the yen has made these investments look even more promising. Nippon Steel can position itself to serve these traditional customers by making parallel foreign investments. It currently has two small ventures in China and small equity investments in downstream processing operations in Southeast Asian countries.[26]

Nippon has no intention of investing in Europe (beyond the significant technical assistance contract it now has with British Steel) or making any additional investments in the United States. Its most recent venture in the American market was the equity position it established in Oregon Steel's purchase of the CF&I Steel Company, and that investment was strategic. Nippon was subject to a dumping suit by Bethlehem Steel for its "head hardened" rail exports. Although it won the suit, Nippon is wary of the possibility of it happening again. Transferring the technology to the United States through Oregon Steel gave Nippon the opportunity to participate simultaneously in the U.S. rail market via "domestic" production and via imports. In this way it spread the risk associated with exchange rate movements and protectionist pressures in the American market.[27]

Competition from Minimills

Japanese integrated producers did not pay much attention to the growing competition from the minimill sector until Tokyo Steel began to challenge their most lucrative markets in the early 1990s. Combined with the realization that the industry as a whole had substantial excess capacity, Tokyo Steel's success focused the attention of the integrated producers on the minimill threat.[28]

The structure of the electric furnace industry in Japan is very different from that in the United States. Independent producers like Tokyo Steel account for about half of Japan's electric furnace production. The other half of the market is satisfied by minimills that are affiliated with Japanese integrated producers. Nippon Steel has linkages with fourteen small steel companies,[29] for example, in a pattern that is typical of the integrated firms. Some of these "captive" minimills produce steel from scrap metal while others simply process semifinished steel produced in other plants. Altogether, Japanese minimills account for over 30 percent of the nation's raw steel production, up from 24 percent in 1980.[30]

Traditionally the integrated companies have regarded their captive minimills as autonomous subsidiaries, letting them produce the less technically sophisticated and more commodity oriented products, such as bars and shapes for the housing and construction industries. This hands-off approach also characterized the way they viewed the competition from the independent minimills—they paid little attention. All this changed when Nippon realized that Tokyo Steel was "invading" its markets, as Nippon executives put it.[31]

Tokyo Steel has a capability of producing about 4 million tons a year and had sales of about $1.5 billion in 1994 (compared to Nippon Steel's 25 million tons and $20 billion in sales). It began to move its product lines into higher value-added markets in 1984 when it opened an H-beam and plate mill. In 1992 Tokyo Steel entered the hot-strip market, and in 1995 it started up a new mill to produce H-beams and sheet piling. It is also considering investments in cold-rolled and galvanizing facilities for 1996. Each of these steps is a direct attack on the integrated producers' most lucrative markets.[32]

Nippon Steel's response to these recent events has been rapid. It competes on the basis of price for all products that overlap, and in some cases this price competition has been severe. In the H-beam market, for instance, Tokyo Steel held a 30 percent market share in the early 1990s, but as steel demand declined, Nippon Steel began to focus on this market through aggressive price competition. By 1994, sluggish demand coupled with Nippon's aggressiveness drove the price of H-beams down by more than half. Nippon Steel and its affiliates succeeded in capturing about 30 percent of the market with Tokyo Steel's share falling below 20 percent.[33]

The impact of this on Tokyo Steel is seen clearly in its fiscal 1994 financial performance. While production fell by only 6 percent, its sales volume (valued in yen) catapulted downward by a dramatic 25 percent, producing the first loss in nine years. Nippon had no intention of retreating from its markets as did the U.S. integrated companies.

Tokyo Steel is the most successful independent EAF producer in Japan. Yet it does not enjoy some of the advantages of EAF companies in the

United States. Scrap prices are 25 percent higher in Japan than in the United States, and Japanese blast furnace producers are very cost efficient. In addition, Tokyo Steel's wage rates and bonus levels are similar to those of their competitors in the integrated sector, and most of the labor practices (job rotation, few job classifications, and few restrictive work practices) are comparable between the two. As a result, based strictly on operating costs Japanese minimills like Tokyo Steel are at about a 10 percent cost *disadvantage*. On a total cost basis, however, minimills compete very well, and low fixed costs make the difference. Some of this can be explained by the low capital costs associated with scrap-based production, but much of the Japanese minimill advantage can be traced back to lean corporate structure and low overhead.[34]

This is why Nippon and other integrated companies have begun to focus on taking disproportionate cuts out of the white collar labor force. Currently, the ratio of white collar to blue collar workers at Nippon is 1:2; after the restructuring it hopes to reach 1:2.5.[35] This compares to Tokyo Steel's ratio of 1:4.[36] Nippon has also utilized outside contractors to a significant extent to reduce its labor costs. More than half the workers in its plants are contract labor, a percentage exceeding that of all the other integrated producers in Japan as well as Tokyo Steel (one third).[37]

Tokyo Steel's advantage in overhead expenses comes from its small corporate bureaucracy. The firm's combined executive and sales offices occupy one floor of a building and consist of about fifty individuals,[38] and the company prides itself on maintaining a fast and flexible decision-making structure. These features make comparisons to firms like Nucor and Oregon Steel very direct; like those firms, Tokyo Steel is focused and driven to compete. Masanari Iketani, President of Tokyo Steel, put it this way, "We have to find a way to survive by ourselves because price competition from the integrated producers is severe. To do this, we need to have a very thin organization, and to be aggressive in new product development and in the use of new technology to make better steel and to make it cheaper. We don't have the luxury of letting up."[39]

This drive is reflected in the characteristics of Tokyo Steel's new plant. Through the use of new technology, an improved layout, and a small cross-trained workforce, the plant is expected to average 5,000 tons of H-beam and sheet piling products per worker per year. This doubles the company's current level of productivity of about 2,500 tons per worker, which already exceeded that of the integrated producers.[40]

The work culture at Tokyo Steel grew out of a situation that existed about twenty years ago when Iketani assumed the presidency. At that time the company was not profitable, and three of its six plants were unable to compete. Faced with the prospect of bankruptcy, Iketani closed the unprofitable plants and dismissed the workers.

Since then, Ikentani has worked hard to reestablish job security. There have been no permanent layoffs in the intervening years, and even in difficult times he has kept on redundant workers until he could find a place for them. The firm's aggressive program for expansion provided those job opportunities. Indeed, most of the employees at the company's new mill will be transfers from existing plants, and there is no intention of replacing the transferred employees.[41]

Because of this history, employees realize that their job security is tied to the firm's exceptional performance, and they know that their work is central to that success. Friendly competition between the plants is encouraged, just as it is at Nucor, by the comparison of monthly costs and production levels across the plants.[42] But behind it all is the realization that unproductive plants can be closed.

Iketani's leadership has been critically important. He knows the technology well, just as do the CEOs at Nucor and Oregon Steel, and he is constantly evaluating ways to do things more efficiently. This sets the tone for the rest of the corporation and has been instrumental in shaping the culture of the firm.

Kaizen: A Built-In Competitive Edge

Achieving a high-performing workplace is an important objective of the minimills and integrated producers discussed in the preceding chapters. These companies are trying to create an environment where employees are motivated to perform well and to focus on improving their performance level. The specific policies and practices adopted to achieve this vary, but clear goals and objectives, open communications, accountability, reward for performance, teamwork, and cross-skilling are common to most of the companies.

In important respects what these companies are trying to accomplish already exists in most Japanese steel companies. The culture at firms like Nippon Steel and Tokyo Steel is conceptualized as *kaizen* or continuous improvement.[43] From the moment employees are hired, they are assigned to an older employee, who serves as trainer and mentor. Often this is the assistant director of the new employee's section. Instruction begins with the norms of behavior and expectations for the way in which employees should approach their work and moves on to specifics about the job itself. There is nothing ambiguous about what is expected. The norm of constant improvement is drummed in, whether one is referring to the individual, the specific job, or the process.[44]

The key to maintaining the commitment to *kaizen* and its strong work culture is trust, which is based on the belief that management is looking out for the best interests of the employees. Job security is fundamental to establishing and maintaining trust because it brings home to the employees the recognition that the company is committed to their long-term welfare. This is reinforced by open communication with the union about most issues, the demonstrated willingness of managers to implement union suggestions where practical, and the autonomy given to teams to make improvements in their workplace.[45]

Any change in the commitment of companies to maintaining lifetime employment would undoubtedly undermine trust and could jeopardize the whole *kaizen* process. This is why large corporations like Nippon work closely with unions to get agreement on the conditions under which workers are transferred to affiliates or encouraged to leave the company. It is also the reason why Nippon has assiduously avoided having to dismiss employees in a manner that would be deemed unfair or inequitable. Smaller companies like Tokyo Steel have been forced to make permanent layoffs when their survival is at stake, but the overall goal has been to tie the well-being of the worker as directly as possible to the well-being of the firm.

Investments in training are part of it, and the training is ongoing. For blue collar workers this takes the form of on-the-job-training primarily. A significant portion of an employee's first year is devoted to training. Improvement in skills throughout one's career is not only encouraged but is expected as part of the work culture. Much of the training that is not job specific would occur off the work site. Continued skill enhancement is an integral aspect of human resource policy in these Japanese companies, and it is more formal and explicit than in American and British steel companies. Lifetime employment most likely creates an additional incentive for this commitment given that companies recognize the employee will be there for many years.

A commitment to cross-skilling reinforces the need for training. Most Japanese steel companies have a policy of rotating blue collar employees in the mills, assigning them different jobs in order to give them greater knowledge about the production process, thereby making them more valuable employees. At Tokyo Steel workers rotate jobs within their work unit on a daily or perhaps more frequent basis just as they do at Co-Steel Sheerness, and there is often movement between adjacent work groups.[46] This increases the flexibility of the workforce, and enhances the *kaizen* process because as an individual's understanding of the production process increases, that person is able to participate more effectively as a member of a process improvement team.

Teamwork is important, and workers think of themselves as an integral part of a team. This is not meant to imply that all teams are equally motivated or successful, but teams are indigenous to the workplace and in most cases they voluntarily engage in process improvement as part of their commitment to the job. Involvement in these "activity groups" (as they are called at Tokyo Steel) or "JK groups" (at Nippon) is extensive. At Tokyo Steel's Okayama plant with an employment level of about 800, there are more than 120 groups.[47] At Nippon Steel, there are more than 3,500.[48] Most employees participate and involvement is strongly encouraged. Most group activities occur during work hours, but to the extent this requires extra work hours, the employees are compensated for the additional effort.

In addition to the results of a specific group's activities, teams serve other valuable purposes. They are tied to the participant's job and work area, and they help to build cooperation. At the same time as they are dealing with issues such as quality, productivity, maintenance, or safety, participants gain a more expansive view of the job and begin to think in broader terms about their job and its connection to the work groups around it. Over time this enables the individual to become a more valuable member of the work group and to think more comprehensively about how to improve the work process.

Financial incentives are not as central in Japanese companies as they are in some American firms. At the plant level, the Japanese argue that workers derive intrinsic rewards from success and improvement, again because this is expected as part of the culture. But there are incentives. Team-based performance is recognized annually in competitions that are judged on a companywide or plantwide basis. Prizes, which include small bonuses and other forms of reward and recognition, are awarded for the most impressive accomplishments. It is generally felt that the visible recognition by one's superiors is key here, not the prospect of a cash bonus; the bonuses are often used to pay for social outings of the group members, reinforcing the group's cohesion.[49]

As discussed in previous chapters, it is the interlocking effect of a number of policies that creates a highly productive work environment, as is especially evident in the Japanese experience. While differences in culture account for some of the variation in the policies and practices across the countries studied, perhaps the most significant difference in Japan, other than emphasis on job security, is the commitment to inculcating the philosophy and culture of the company in workers from the moment they walk in the door. This has a lot to do with the willingness of workers to focus on continuous improvement as part of their jobs, and it reduces the need for explicit financial incentives and special programs to bring about improvements in quality and productivity.

Summing Up

The Japanese steel industry differs from that in the United States or Britain in a variety of ways. The company unions have worked more closely with management in recent decades, and the unions have permitted practices such as contracting out, job rotation, minimum job classifications, and job descriptions that are based on flexibility to a greater extent. All these factors have enabled management to organize production and deploy workers in an efficient manner. In recent years the integrated companies in the United States and the U.K. have been able to negotiate many of these advantages with their unions, but not to the degree they have in Japan.

The Japanese integrated producers are going through many of the same reorganization and rationalization processes that were experienced by their counterparts in the United States and Europe a decade ago, and they are contracting out jobs in the mills to a much greater extent. But the urgency of change is still not as great. The flow of red ink that propelled British Steel, United States Steel, and others is much smaller in Japan, and it is likely that Japanese companies will return to profitability in the near future. As a result the speed of restructuring has been slower. Still, the pressure is there.

Traditional Japanese management practices have embraced many aspects of what it takes to create high-performing workplaces. The challenge facing the integrated companies in Japan is to decide how fast and to what extent they want to create a thinner organizational structure and empower decision makers at lower levels. Reorganization and decentralization will come to pass in firms like Nippon Steel, but the pace will depend upon the intensity of competition from steelmakers in other Asian nations and from minimills like Tokyo Steel.

Tokyo Steel has indicated that it has every intention of continuing to move into higher value-added products, and as technological advances permit its assault on the markets of the integrated producers will escalate. This pattern of competition is similar to what has occurred in the rest of the world. Technological change can open doors to opportunity. This has been true for minimills, and it has been true for integrated manufacturers. In the next chapter we turn to the technology side once again, where we will find the staging ground for competition on the near horizon and beyond and explore the implications of technology drivers for the future structure of the steel industry.

9

Converging Technologies

The vitality of the American steel industry and its structure in terms of the kinds of firms that characterize the industry as we know it today result from the fact that individuals seized opportunities available to them. Invariably those opportunities were in the form of extraordinary profits to be realized by taking full advantage of existing technology or by pushing the technology frontier. The forces driving that opportunity were those of the marketplace, whether the market involved was steel or that for the materials needed to produce steel. Overall what we have observed is competition, foreign or domestic, forcing downward pressure on prices in product markets or moderating price increases, which created a profit squeeze that caused firms to reexamine current practice. These conditions were exploited by those bold enough to take the risks involved.

The key players in this contest achieved their successes by carving out advantage in particular markets and marshaling the resources at their command to sustain their position. While the ability to identify opportunities often required deep insight on the technology side, resources had to be organized and people had to be motivated to succeed in order to make the most of these opportunities.

The steel industry is going through a technological revolution that is changing the way steel is produced and the structure of the industry itself. Firms are now rushing into the market for flat-rolled products with a technology proved viable by Nucor or variants of that technology—thin-slab casting. As described in Chapter 4, by the turn of this century as much as 18 to 20 million tons of new capacity in electric furnaces and thin-slab casting may be up and running. That is in the order of 40 to 45 percent of the current level of domestic shipments in this product line.

Before Nucor's entry in this market, flat-rolled steel had been the exclusive domain of large integrated producers. Forced out of low value-added product markets by low-cost minimills and stiff competition from abroad, integrated firms made a comeback to prosperity in higher valued, flat-rolled products. They will not back off easily this time.

In this chapter, we assess this critical juncture by characterizing the technological forces shaping this industry's future course. In the process, we will describe a remarkable convergence of steelmaking technologies that is on the near horizon—future technologies that will shape competition and help to sort out a new set of industry winners and losers.

Technology Drivers

Capital cost advantage figured prominently in initial minimill successes, and it figures prominently today in their efforts to further reduce the market share of integrated producers in flat-rolled products. The production costs presented in Chapter 3 make this point (see Tables 3.1 and 3.2). Conventional integrated plants, geared to produce 4 million tons per year, require enormous capital expenditures. The fixed capital tied up in making steel by this traditional method—from that required in processing material and energy inputs to that embodied in ladle metallurgy and vacuum degassing (see Figure 9.1 below)—is estimated to be \$900 to \$1,500 per ton of annual capacity. Continuous casting in conventional thick slabs and

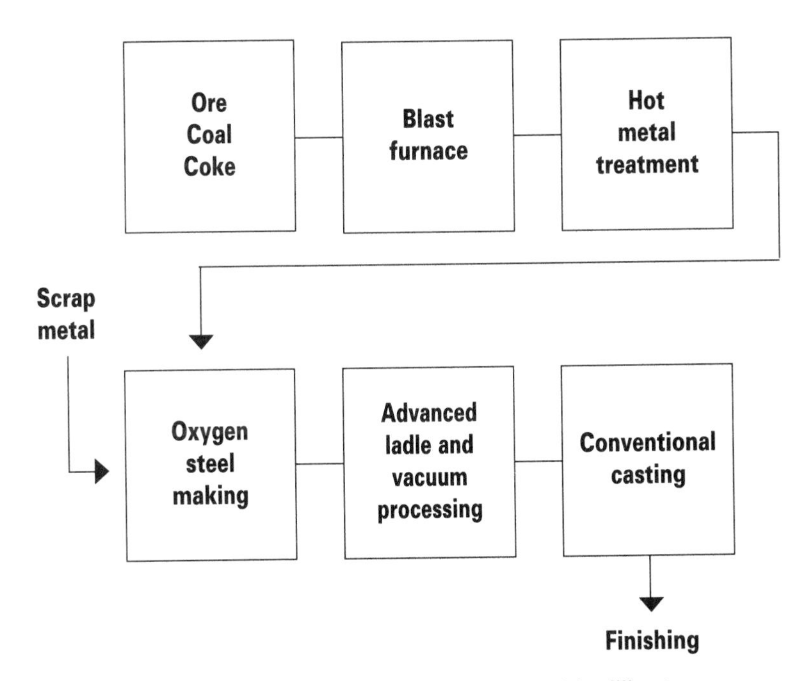

Figure 9.1. Stages of integrated production: capacity of 4 million tons per year.
Source: R. J. Fruehan, H. W. Paxton, F. Giarratani, and L. Lave. *Future Steelmaking Industry and Its Technologies.* Idaho National Engineering Laboratory, Report INEL–95/0046 (December 1994).

the associated rolling capacity bump these figures up by another $300 per ton. In contrast, the traditional minimill that processes scrap in an EAF furnace knocks the capital expenditure for steelmaking down to only $140 to $170 per ton of annual capacity. Casting and rolling operations would add only $180 per ton to that total.[1]

The technological advantage of scrap-based producers that is reflected in these cost estimates stems primarily from the elimination of blast furnace and coke operations, which dramatically reduced the capital costs per ton of production. At the same time, this lowered an immense barrier to entry because production was now viable on a small scale.

The minimills also took advantage of very favorable conditions in the market for scrap metal, so that initially their operating costs were held down by low prices for this key input. Recently, however, much of the competitive advantage from this source has been chipped away as scrap metal prices have ratcheted upward in response to increased demand. This change in the market now helps to define the frontier of opportunity.

To compete in flat-rolled products, where market share depends crucially on the quality of the sheets produced, low-residual scrap metal or mixed charges that include scrap as well as virgin iron have to be used. At current scrap prices, the higher cost of a furnace charge based exclusively on low-residual scrap metal substantially reduces the minimill advantage from capital cost savings. The cost of a mixed charge, however, allows them to remain competitive if pure scrap substitutes are available at a reasonable price.

These facts drive the incentive for the adoption and change of technology in both sectors of the industry, integrated producers and minimills. To protect their position in the heart of the market for flat-rolled products, integrated producers must develop processes that reduce their capital costs and make production on a smaller scale viable. To penetrate that market, the new thin-slab minimills will have to control the quality of their raw materials at the same time as they restrain material costs.

The Integrated Sector

Integrated producers have plenty of reason to seek the bulk of the capital savings they need in coke-making and iron-making operations. From an engineering perspective, the blast furnace is highly efficient for iron making, and its productivity and fuel rate have been improved significantly in recent years. But besides being highly capital intensive, the blast furnace requires coke. This fuel is made by transforming coal into a more potent material with high energy content and fewer impurities, a transformation that requires heating coal in ovens. In the process, large amounts of residual gases are produced, distilled into by-products, and sold. There are also very significant gaseous leakages from the oven batteries, however, and these emissions are toxic as well as carcinogenic. The water used to quench coke is also contaminated.

The coke industry is under the intense scrutiny of the U.S. Environmental Protection Agency, and the Clean Air Act of 1990 set stringent standards for benzene emissions and coal tars in coke-making operations. Companies are now being pushed to the limit of technology in controlling these pollutants. Abatement costs have been driven up dramatically, which in turn has translated into increases in the cost of coke. Moreover, further increases are threatened in the near term if risk analysis shows that emission levels in coke making are still too high. In that event, the design of coke ovens may have to change fundamentally. This would come only after major new investments in oven batteries that would sap investment capital from other crucial stages of production and drive production costs further upward. One implication of all this is that market analysts now predict a coke shortage in the near future.[2]

With the price of coke rising and expected to rise even more, steelmakers have sought ways to reduce their dependence on this energy source. Currently, many major steel plants are installing "pulverized coal injection" into blast furnaces as a partial substitute for coke.[3] A few use natural gas for the same purpose. The need for greater efficiency in blast furnaces (i.e., to get more hot metal per ton of energy input) has also focused attention on the efficiency of oxygen use so that combining coal injection and large amounts of oxygen can further diminish the need for coke.

This kind of evolutionary change will likely dominate technological developments in the industry over the next few years. But beyond that there will be a drive to virtually eliminate the use of coke in many plants by adopting processes that require only coal and no coke to convert iron ore to crude pig iron. One such process, Corex, is commercially available now,[4] and several other processes are currently under development and likely to become commercially viable in ten to twenty years.[5]

The very same kinds of incentives are characteristic of steelmaking operations. Capital costs must be reduced, furnace efficiency enhanced, and flexibility in the mix of inputs used (e.g., virgin iron versus scrap metal in the charge) increased. Technical developments are under way in each of these areas, and as a consequence the alternatives for converting pig iron to steel are likely to expand in the near future. All this implies that capital costs for integrated production will come down, and with this decline will come a reduction in the minimum efficient size of integrated plants.

Importantly, as coke operations are eliminated and oxygen steelmaking is reduced in scale, it will become viable to wed new types of casters to integrated operations—casters that can also operate efficiently on a smaller scale. Thin-slab or mid-thickness casters could be used. Strip casters that produce sheets of only one-tenth of an inch in thickness and roll them directly into coils are also very close to commercial use.[6] The use of near net shape casters will become more commonplace. On the basis of current technology or technology that has high promise of commercialization, one can envision efficient ore-based plants with a capacity of only one or two million tons per year that produce coils or beams (from shape casters) of very high-quality steel. In effect, the "mini" integrated plant will have been born (see Figure 9.2).

The Minimill Sector

Rising scrap metal prices threaten the profitability of minimills across the board, and the availability and price of low-residual scrap metal limit the ability of minimill producers to penetrate the market for high value-added products. With substantial new thin-slab capacity coming on line, some observers see a "blood bath" in the making. The integrated firms know that they are at risk, but they believe that the new minimill entrants are at risk too. High prices for good quality scrap could very well sap the profits from

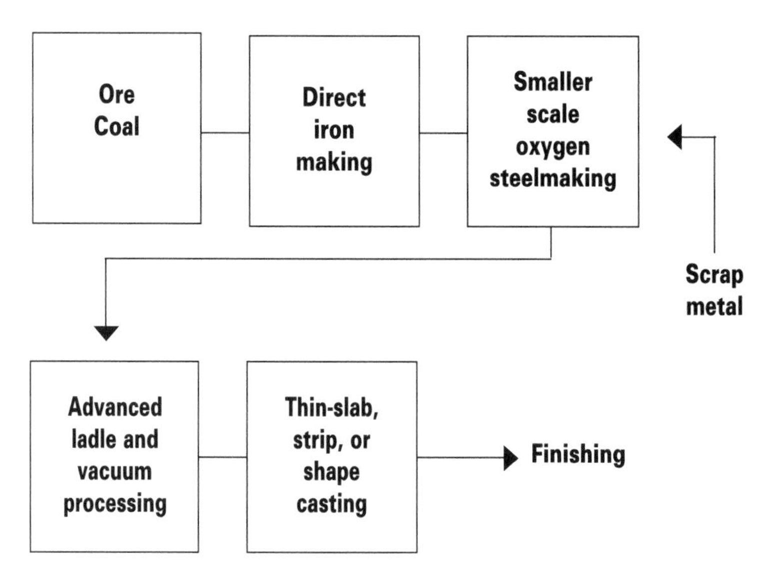

Figure 9.2. **The "mini" integrated steel mill: capacity of 1–2 million tons per year.**
Source: R. J. Fruehan, H. W. Paxton, F. Giarratani, and L. Lave. *Future Steelmaking Industry and Its Technologies.* Idaho National Engineering Laboratory, Report INEL–95/0046 (December 1994).

their new rivals. The ensuing struggle could mean a new round of plant closures for the integrated sector, but could also reveal some new thin-slab ventures as poorly planned and poorly timed.

It is far from clear which sector, integrated or minimill, will bear the major brunt of failure in this contest. The scrap metal market is being tested as never before in the post–World War II period. Much turns on whether the increase in scrap demand that will come along with the new thin-slab capacity is met by a comparable expansion of the supply of scrap metal. Also critical is the rate at which scrap demand actually does increase and whether or not effective scrap substitutes are available.

There are three basic sources of supply for scrap metal: (1) trimmings and spillage in steelmaking plants, or "home" scrap, (2) scrap generated in the production of steel-intensive goods like automobiles and tractors, or "prompt" industrial scrap, and (3) "obsolete," post-consumer scrap. The highest quality steels require large percentages of home or prompt industrial scrap, but these commodities are limited and their availability is decreasing due to the implementation of continuous casting and improved manufacturing of steel-intensive products.[7]

The stock of obsolete scrap is huge. Some estimates range at up to one billion tons in the United States alone. However, only a fraction of this can be recovered. One can conceive of the steel-intensive products of yesterday (say, twenty years past) as being "obsolete" now and available for use as scrap in steelmaking. Using that as a base, historical recovery rates for scrap metal have usually been estimated to be in the neighborhood of 30 to 35 percent. But it would be a mistake to believe that these historical averages will accurately represent future recovery rates. With scrap prices at historical highs and demand projected to grow, there is ample profit incentive for scrap distributors to seek ways of extracting more scrap and possibly more low-residual scrap from the existing stock. It is conceivable too that government intervention could mandate recycling. In the longer term, one might even envision that on a large scale, steel products for end users could be engineered with scrap recovery in mind. The point is that the elasticity of obsolete scrap supply in the long term may be larger than one could reasonably predict from the historical data. With new opportunities for profit there are certain to be new ideas for scrap recovery, and recovery rates could move to 50 percent or more in the long term.

The use of low-residual scrap will not expand in strict proportion to the increase in capacity that is on the horizon for thin-slab plants simply because there are insufficient supplies. Rising scrap prices will pressure a broad cross section of steel firms to seek scrap substitutes for their furnace charge. Adding some form of reduced ore, such as DRI, HBI, or pig iron, dilutes the impurities in obsolete scrap and results in the steel chemistries required for flat-rolled products. The same result can be achieved by mixing liquid hot metal with scrap.

In some steel plants this will require a new mind-set. Instead of rigidly seeking a fixed ratio of inputs in the furnace charge, financial success may well depend on a firm's ability to vary input ratios without sacrificing consistency in terms of product characteristics. Different blends of materials will have to be used to meet the quality standards required for various markets at the lowest possible cost. And there will be an even higher premium on steelmaking know-how as a result of this.

The locations of the announced thin-slab plants suggest that at least some of the entrepreneurs involved understand full well the need for this kind of flexibility. The sites chosen preserve options for water and rail transport of scrap metal; they are not limited to one or the other. Most certainly this was a key factor in the location decisions. But scrap substitutes use exactly the same transport options, so the new or proposed plant locations provide access to those inputs too. People who can convince investors to commit up to half a billion dollars to a new steel venture are not likely to be myopic when it comes to a decision that will determine whether the ledger is written in red or black ink.

Some scrap-based minimills that have already expanded into other raw materials may well serve as prototypes for the new market entrants. Nucor is a case in point with its iron carbide plant in Trinidad. Another is Georgetown Steel in Georgetown, South Carolina, where large amounts of direct reduced iron are produced on-site and used along with scrap metal in electric furnace steel production.[8] Still another is British Steel's Tuscaloosa, Alabama plant, which will have its own source of direct reduced iron from two Midrex DRI units that are being relocated from Scotland to Mobile, Alabama.

Schematically the advanced scrap-based plants that will characterize the frontier of steelmaking in electric furnaces can be represented as in Figure 9.3. The plants may include processes to produce clean virgin iron to be used with scrap metal in electric arc or other types of melting furnaces. Firms that absorb this additional capital cost burden will be integrating backward toward raw material supplies in order to secure access to higher-quality markets.

A comparison of Figure 9.3 with the schematic representing mini-integrated production (Figure 9.2) reveals a convergence of technology in the industry. Ore-based and scrap-based production will find common ground in plants with a capacity of 1 to 2 million tons per year. The furnaces involved will accept and help optimize a charge with flexible components,

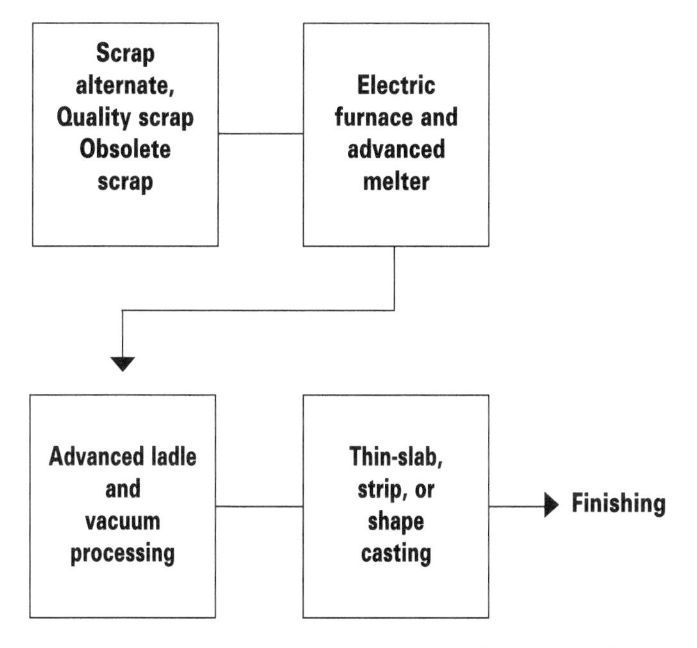

Figure 9.3. Advanced scrap-based production: capacity of 1–2 million tons per year.
Source: R. J. Fruehan, H. W. Paxton, F. Giarratani, and L. Lave. *Future Steelmaking Industry and Its Technologies.* Idaho National Engineering Laboratory, Report INEL–95/0046 (December 1994).

and they will be combined with the use of advanced ladle and vacuum processing to control the final steel chemistries to closer tolerances. Together, these scrap and ore-based prototypes define what might well become the "midsized core" of plants designed to produce low- and medium-quality products for markets in flat-rolled steel.

The lines of demarcation between the integrated sector and minimills are being blurred in yet another dimension. With a few exceptions, integrated producers in the United States gave up EAF production in the 1970s and 1980s largely because they could not compete with the independent scrap-based mills that were coming on-line to produce simple long products. At that time, the emerging minimills were just too efficient and too dynamic, while many of the integrated firms were mired in past failure. However, there is a turnaround in the making today. Far from rolling over in defeat, integrated producers are returning to EAF production and following the Nucor lead into thin-slab casting. Gallatin Steel is one example of this. That plant, located in Kentucky, is a joint venture involving Dofasco Inc., an integrated producer based in Hamilton,

Ontario, and Co-Steel, a Canadian minimill. Gallatin's initial scale is just over a million tons per year, and they are planning for expansion.[9] LTV, in a joint venture with British Steel and Sumitomo Metal Industries, is building a 2 million ton plant (Trico Steel) based on EAF thin-slab casting technology.[10] Both plants will secure a nonunion workforce, and will use less than one man-hour of labor per ton of steel. Neither of these ventures could be characterized as "mini" in any meaningful sense.

We can expect convergence in the organization and culture of the workplace too. The integrated producers will use labor more flexibly. They will do this within a nonunion environment, like Gallatin Steel and Trico Steel, or within the limits of an organized workforce, but they will do it nevertheless. Without more flexibility they cannot make the most of their technology and cannot get the financial returns they need to thrive. A good example of this is in the area of maintenance, which represents one of the emerging areas of cost consciousness in steelmaking. In many nonunion plants, maintenance is carried out by operating personnel, whereas union plants rely on special crafts workers for maintenance operations. The evidence on the shop floor suggests clearly that the use of operating personnel in this capacity not only reduces labor costs by utilizing workers more efficiently, but that this practice is also associated with efficiency gains in terms of the percentage of time that machinery is on-line and working properly.

A New Margin for Domestic Competition: Specialty Steels

The competitive struggle that we have described has primarily involved battles in particular carbon steel product lines. In total, nearly 120 million tons of carbon steel were consumed in the United States in 1994.[11] The market for specialty steels—stainless steel and alloy tool steel—is much smaller: In 1994, consumption amounted to little more than 2.5 million tons.[12] Like minimills, specialty steel producers rely on electric arc furnaces charged with scrap metal. They also operate on a small scale, like minimills, but they go one step further to get the properties they need in finished products by adding alloys like chromium and nickel in the electric furnace or ladle.

By far the biggest competitive threat to U.S. specialty steel producers has come from imports. As the industry restructured, the strategic moves by integrated firms and minimills were in carbon steels. But demand for specialty steel in the United States has been robust. Over the period 1989–94, apparent consumption in this market increased by 700,000 tons (an increase of 39 percent).[13] The profit margins have been high, perhaps 25 percent or more. This combination of strong demand and good profits has attracted the attention of foreign producers. Indeed, more than half the gain in the consumption of these products in the last five years has been satisfied by imports, and the market share of imports in specialty steel products is now 34 percent.[14]

Given the new competitive strength of U.S. carbon steel producers, it was only a matter of time before they began to look to specialty steel markets as fertile new ground. In the last few years Nucor has begun to produce limited quantities of specialty steel at its Crawfordsville plant, and its recent investments suggest more emphasis in this market in the very near term. Lukens Steel, another large scrap-based producer of carbon steels, has recently purchased Washington Steel, a stainless producer, and Lukens has also invested in technology that will give its Coatsville plant the capacity to produce stainless products. Previously that plant was limited to the production of carbon steel plate.

As these firms learn more about producing stainless products, it is a sure bet that they will move from the low end of the quality distribution to ever higher quality products. Also, if the prediction of a new midsized core is realized, there could well be entry at the top end of the quality spectrum too. Right now, Kawasaki Steel in Japan is producing stainless steel from an ore-based process that allows it to smelt chromite ore directly in the steelmaking furnace.

The advent of stip casting would also intensify the competitive pressure in specialty steel markets. As a rule it is easier to cast stainless steels as strip than it is to cast carbon steels in this form. Thus if any of the new strip casting technologies that are now in the pilot stage materialize as commercial ventures, they are likely to move first into stainless product lines.[15] Based on the technologies involved, plants with a production capacity in the range of 250,000 to 500,000 tons per year of flat-rolled steel may well be viable with this combination.[16]

Strategies for Technological Development

In-house research and development activity has not played a central role in the American steel industry's turnaround. With rare exception, minimills have shunned research. Nucor professes to be proud of the fact that it has no research department. Instead its plants serve as research laboratories for the implementation or improvement of technology developed elsewhere. A few other minimills such as Chaparral Steel in Texas, North Star Steel in Minnesota, and Georgetown Steel in South Carolina have carried out research programs with some success, but in all cases the activities are limited in scope and small, as compared to domestic or international integrated producers. Rather, as exemplified by Nucor, minimills have sought success in the marketplace by an aggressive strategy aimed at purchasing, implementing, and perfecting new technologies as they become available on the world market.

The remarkable gains achieved by U.S. integrated producers also had little to do with advantage from their research activities. In fact, these producers have been shedding research personnel and whole research departments for the last twenty years. When their profitability began to suffer in the mid–1970s research staffs were among the first to go. For example, over 1,800 people were employed in research at USS in the early 1970s, whereas fewer than 150 are employed there in a research capacity today.

Instead, there has been a partitioning of research activities. Applied work remains the province of the firms, with personnel engaged in research on the plant floor or in interface with customers. Basic research has largely been shifted to cooperative ventures where industry, government, and universities become partners, and major funding often comes from the government side.[17] Government-industry cooperation in research here has counterparts in Europe and Japan. The importance attached to such activities is reflected in the exception granted to government subsidies of this sort in international trade agreements.[18]

The other source of advantage in technology development and implementation has come from joint ventures, especially those between U.S. firms and foreign partners. When American firms were playing catch-up in the 1980s, many took the highly successful route of identifying a market niche that could be exploited with technology that was already in place and being used by foreign producers on the technology frontier. Alternatively, they sought new technology from firms specializing in the manufacture of steelmaking equipment, and they entered new markets as a result, sometimes in partnership with the equipment suppliers. Indeed, in plant visits here and abroad, one cannot help but notice the incredibly open nature of production facilities. It seems that trade secrets are few and far between in this industry. The information flows, both interfirm and international, are well developed and well used.

Today, however, it is American firms who are helping to define the technology frontier in many areas. They are not playing catch-up any longer. The technological advances that will be necessary to bring about the midsized plants we have described may well depend on American initiatives, not those in Europe or Japan. The profit incentives in these developments are greatest here, because the competition involved is most intense here. Because of this, the forward-looking firm, whether minimill or integrated, may find that in the coming years a technology strategy that takes the initiative for development away from partners and puts it closer to home may be the best route to a breakthrough that can establish market advantage and maintain it.

Summing Up

It is not possible to predict with accuracy whether the aspirations of the new scrap-based thin-slab producers will be realized or frustrated. The expected surge in flat-rolled capacity inevitably leads to concern that a repeat of the 1980s may be in the offing: A decade ago, this industry was burdened by excess capacity and plants were closing because they were no longer viable. Now, up to 20 million tons of capacity are to be added to a market that is presently in the neighborhood of 50 million tons per year. (In 1993, net domestic shipments of flat-rolled products were 44 million tons and imports were just over 6 million tons.)[19]

Few in the industry should expect extraordinary growth in domestic consumption; 1 or 2 percent per year would be a prudent guess. As the new thin-slab capacity comes on-line, some imported steel sheets and strip will be replaced by output from lower-cost American suppliers. And American steelmakers may finally begin to exploit the advantage that they enjoy on world markets and export steel abroad in significant quantities. Indeed, at present they have the very formidable advantage of low-cost producers whose selling price is reckoned in a weak currency on international markets. American steel should look awfully good to European and Asian manufacturers.

Will all the new thin-slab capacity be absorbed in these ways? Probably not. Inevitably, it seems, those integrated producers that focused their restructuring on the lower end of the flat-rolled market are in for some stiff domestic competition. They in turn may refocus on higher value-added products in the flat-rolled market,[20] so we should expect competitive pressure to be turned up generally. The closure or downsizing of some operations may well be a part of this. But the restructuring may also speed up the advent of mini integrated plants, and set the stage for a redefinition of the American steel industry comparable to that forced by the successes of modern minimills.

10

Achieving and Sustaining Competitiveness

We have argued that the path to competitiveness for any company is to be found in its people and its technology. Leadership is needed to establish the vision of a company's goals in particular markets and to chart the course of action required to serve those markets. Communication, trust, training, empowerment, reward for performance, appropriate technology, customer focus, quality, and cost reduction are what goes into creating a success. These elements can support and complement one another, and they can coalesce to create a corporate culture that leads to behavior, actions, and decisions that position a firm for success.

Technology has been one of the driving forces in the transformation of the American steel industry, and our findings point to two lessons in this: First, technology can provide strategic advantage in particular markets. Second, it sets the stage for those who understand that coordinating investments in the workplace with investments in machines can lead to real change and real competitive advantage. Successful managers learn both these lessons extremely well.

The technology frontier that we described in Chapter 9 defines opportunities for innovation that exist today and opportunities that are on the near horizon. The bold will be stepping out onto that frontier, and others will be drawn to it inevitably. In this chapter we look at the characteristics of the firms that will be in the forefront of this evolution, in order to examine the management practices that will be required to lead and succeed. We shall also summarize a few of the most central lessons that we have drawn, with the objective of demonstrating that the principles underlying success in global competition generalize widely.

The Vanguard of Change

Our analysis of restructuring in the steel industry leads to definite conclusions about the firms that will be in the vanguard of market competition.

They will have the day-to-day advantages of the high-performance firms we have described and the dynamic advantages that come naturally with that culture.

New firms, or for that matter firms that have been reconstituted after bankruptcy, have a distinct advantage in that the concept of "focusing" investments to achieve market position is, almost by definition, what they are about. Even if the first plants created in the new midsized technology core arise from restructuring in one of the traditional integrated firms, the size and scope of the decisions involved are such that a comprehensive investment strategy will be a prerequisite for action. Venture capital of the magnitude required here is not available from the money markets if firm objectives and technology strategy are not well defined and well coordinated. This necessary condition for success is likely to be satisfied by any firm that attempts to be a technology leader.

The technological advantages realized by a company, however, will be short-lived if its leaders do not recognize and respond to the inevitable pressure brought on by market entry. In the steel industry and many other industries, technology is readily available to competitors worldwide, and a company's long-run competitive edge will therefore depend on the skill of its workforce and its ability to derive advantage from the way it manages the process of change. Ultimately it will be the human side of the enterprise that will determine the ability of the firm to stay ahead of its competitors.

There are certainly very profitable firms in the steel industry today who adhere rigidly to traditional management practices. But we have found that truly outstanding performance is often tied to management strategies that are highly integrated and that cut across all facets of the business to get employees working toward a common objective. These strategies build toward the overriding goal of striving to lower cost and improve quality as an ongoing part of the production process. But they also impart an advantage to firms by allowing them to make the most of the technology they use.

The widespread adoption of continuous casting helped to change the work environment in steel mills fundamentally. With the buffer of ingot stocks removed, the efficiency of operations and the productivity of the machines came to depend upon the ability of the workforce to maintain the flow of production. Firms could maximize that benefit by doing three things: (1) They had to decentralize decision making. (2) They had to invest in the knowledge base of their workers. (3) They had to invest in organizational change.

Put simply, the cornerstone of successful competition is management that takes a very broad view of the concept of investment. Changes in technology have to be coordinated with changes in the workplace, and both require the commitment of time and money with the objective of reaping benefit over the long haul. Just as investment in continuous casting enhanced the value of investment in technologies that helped control the chemistry of steels to finer tolerance, investment in continuous casting also bred investment in changes in the work environment. High-performance workplaces are those that have identified and capitalized on this kind of complementarity in production.

Apart from the difference in technology, the main distinguishing features of minimills, as compared to the larger integrated producers, are their small plant size and the streamlined character of their organization. The leading performers among the minimills are also predominately nonunion. These differences helped speed the organizational changes that are required to make maximally effective use of today's technology.

The fewer the number of employees, the easier it is to ensure that they are all pulling in the same direction. It is also easier to establish a shared vision for the company, to ensure that workers know what is expected of them, and to help them see how their actions affect the company's performance. The same benefits arise as the number of organizational layers declines. In addition, as the number of decision makers falls and responsibility is decentralized, decision making can be speeded up. A nonunion workplace means that managers can talk directly to the workers, without having to involve an intermediary, thereby improving communication. In addition, cross-training and greater flexibility in the way workers are utilized improves efficiency and lowers costs.

There is no reason why these benefits cannot be realized in larger plants or in unionized environments, as evidenced by AK Steel and British Steel. But as plant size grows and the layers of organization increase, rapid and consistent communication becomes more difficult and the messages from the top down are more likely to be inconsistent. The net result is that the vision for the company or the plant is not as clear, and the behaviors growing out of the culture are not as consistent. All this makes it more difficult to ensure that everybody is working toward the same goals. To the extent this occurs, acceptance of change and the pace of change may be slower in these larger work environments. The leaders in the new midsized plants we have described could learn these lessons and prosper.

Before the upheaval in the 1980s, steel industry management was often unresponsive to customers and workers, and it was slow to adopt new innovations. The most competitive companies and plants today have achieved their success because of a different approach to management. Their workers are engaged in activities that promote firm success because the workers are themselves motivated to succeed, and they want to gain the technical know-how that will allow them to do so. Self-improvement through skill acquisition is ingrained, as is the idea that process improvement must be continuous. The outcomes of this are reckoned in terms of advances in quality, productivity, and customer satisfaction—as well as the bottom line.

The management in the better-performing companies understands that it is the interlocking, consistent, and supportive aspects of managerial decisions that create the culture of the workplace, which in turn drives the behavior of the employees. They have to invest in building a corporate culture and an organizational structure that encourages continuous improvement and decentralizes decision making.

For the companies at the heart of our study, striving to become the low-cost producer in their relevant markets is a well-accepted objective, and they have put policies in place to accomplish this. Investments in technology are made to support movement into higher value-added market niches in order to secure higher profit margins. Training is emphasized to ensure that employees have the requisite skills to support the transfer and use of more advanced technologies and to increase the flexibility of the workforce. In most companies, particularly the minimills, incentive systems are in place to motivate employees to improve their work, and the motivation is an ongoing part of day-to-day operations. Openness and communication are emphasized to improve the decision making process, and responsiveness to customers is stressed as a part of everybody's job.

Firms with these characteristics sustain the advantage that comes from strategic capital investments in the sense that the return on those investments is enhanced by the organization of the workplace and the culture of the work environment. Just as important, the payoffs from investment in the organizational changes and management practices that define the culture of a high-performance steel mill are enhanced by the technology employed in modern mills. Today's technology demands a flexible and knowledgeable labor force, and an organizational structure that supports effective decision making at the technology interface.

Dynamic Advantage

Each of the firms we have profiled has succeeded in gaining access to new markets by taking market share away from their rivals. Some, like Nucor, have been extraordinarily aggressive in this respect. Their active expansion programs have an element of learning-by-doing, for having once pushed into a new market by implementing a new technology, the costs of doing that again—in another market at another time—are inevitably reduced. Locational decisions will be refined, less time will be taken from planning to start-up, and savings will be realized in the costs of establishing and training a new workforce. As important, the growth achieved opens new opportunities for advancement to workers at all levels and this further establishes the linkage between the company's well-being and that of its employees. In turn, this encourages investment by employees in terms of time and energy, which yields profit for the firm.

So in an important way the success of these firms is tied to their ability to understand and anticipate industry change and then respond to it aggressively. The significance of this is well captured in a recent book by Tom Peters.[1] His basic premise is that because the world is changing rapidly, firms cannot stand still, and an overriding objective of management must be to recognize this dynamic and develop their organization's capacity to prosper within it. He advocates building a corporate culture that helps to break down internal resistance to change based on traditional ways of thinking.

The firms at the forefront of the steel industry's transformation have these characteristics. They could also be described as "learning organizations."[2] An important aspect of their success lies in the fact that knowledge acquisition and knowledge utilization are expected in these firms, continuous improvement is the norm, and change is something that they embrace rather than fear.

Creating such an organization requires that policies, practices, and goals reinforce one another. Knowledge acquisition has to be facilitated through human resource policies that encourage education and training, provide employees the opportunity to learn and apply new knowledge, and reward them for doing so. Mechanisms must be in place to seek out new knowledge and bring it back to the company. And there must be internal processes to share the information within the organization, assess it for the opportunities it presents, and act on it where appropriate.

Such an organization works best if cross-functional communication is the norm and entrepreneurial behavior is encouraged. Change and innovation have to be seen as the normal way of doing business. This requires a corporate culture that values these behaviors and leadership that stresses their importance. Companies with employees who can operate effectively in less structured environments, who have good teamwork and team-building skills, and who are more entrepreneurial than their predecessors will excel in the years ahead. Corporations mired in tradition and slow to react to change will not prosper.

Again, Nucor provides a good example of a dynamic learning environment. The company worked closely with vendors in selecting the equipment for the Crawfordsville start-up, and the learning from Crawfordsville and from other Nucor plants, coupled with the additional experiences of the vendors, was incorporated into the equipment selection, engineering, and construction of the Hickman plant. Likewise, the learning from

Hickman, other Nucor plants, and vendors will influence the changes incorporated into the South Carolina start-up. Nucor also reaches outside the firm to meet its staffing needs for new start-ups, and this brings additional technical experience into the plant. The company learns from its customers as well. Partnering with customers provided the impetus for the Hickman plant to produce more grades of steel and to roll wider and thinner than it had originally envisioned. The learning environment produces tangible benefits. Start-up costs at Hickman were almost $30 million less than at Crawfordsville, and the Hickman plant is capable of serving a wider variety of markets than originally anticipated.

Finally, the culture within Nucor nurtures ongoing improvements as the workforce constantly strives to do better. Rodney Mott was Vice President and General Manager at the Hickman plant and is now in a similar position at the South Carolina start-up. He explained this aspect of Nucor's culture in terms of a competitive spirit, "It's a bit of an ego thing, the feeling of accomplishment from constantly improving performance. Money is not the primary motivator, our employees want to excel; we benchmark ourselves against other Nucor plants, and each of our crews strives to be the best in our plant. This is part of the Nucor culture."[3]

We do not want to leave the impression that Nucor is unique in the features we have described. It may be the best comprehensive example, but many of the firms we have visited, some not mentioned here at all, share in various aspects of this profile. The key to all this is lodged with management. Top management sets the vision and develops the strategies to get there, and its policies and practices establish the culture that guides the behavior of all employees. Success depends upon the soundness of the strategies employed and the skill involved in their execution.

The Importance of Competition

Our study illustrates the importance of competition in determining a company's or an industry's level of attention and commitment to ongoing improvements in productivity, quality, and customer responsiveness objectives. It took severe competition from foreign imports in the early 1980s and the growing presence of the domestic minimills (and the large losses that ensued), to convince the U.S. integrated firms to restructure. And since that time the need to remain competitive has not waned as the electric furnace producers continue to chip away at the integrateds' markets. Likewise, the need to restructure British Steel was driven to a great extent by its large losses, the curtailment of its government subsidies, and foreign competition. Recently, the integrated producers in Japan too have been forced to alter the traditional way they have carried out their business, in part because of foreign competition in their domestic as well as their export markets.

As we have explained, these pressures will not abate in the future. In fact, they will accelerate. The costs of market entry have been lowered in the steel industry, which makes it likely that more new companies will be formed and existing companies will expand to enter high value-added markets. Market success has to be defined within this competitive context.

Where market niches appear to offer the prospect for future profits entry will occur, putting added pressure on those already in these markets. Examples include the decision of Broken Hill Proprietary Company's BHP Steel to build a mill in the Pacific Northwest to produce cold-rolled coated steel. This is part of BHP's global strategy to supply the Pacific Rim. Recently the company completed plants in Malaysia, Vietnam, Indonesia, and Brunei, and it plans to invest in other parts of the region in the future.[4] Certainly, competition in Asia will heat up. British Steel has targeted that area for growth and investment,[5] as has Nippon,[6] but so far these companies have not been as aggressive as has BHP.

Closer to home, additional competition could come in unexpected ways. British Steel is making a significant investment in its Tuscaloosa Steel facility to turn it into a full-fledged minimill by adding a melt shop and modernizing the rolling mill.[7] Brian Moffit, Chairman of British Steel, elaborated, "The expansion of Tuscaloosa's operations is designed to create an internationally competitive minimill to produce quality plate in coil and cut length form. Tuscaloosa's objective is to become the most cost-effective supplier in the U.S. plate market, and this investment is an important step forward in British Steel's overseas plans."[8]

Or take the example of the joint venture of British Steel, Sumitomo Metal Industries, and the LTV Corporation to build a flat-rolled minimill in Decatur, Alabama, capable of producing two million tons annually. Trico Steel Company, as it is called, is a unique partnership between three integrated steel companies. If successful it could serve as a model for expansion into other parts of the world, principally Asia.[9]

Competition in the steel industry will accelerate in the years ahead. New technologies seem to be developing at a more rapid rate, and the diffusion of these technologies internationally is facilitated by vendors who sell their equipment globally. As there are no natural barriers to this process, the market advantages given to the company that adopts a technology first can be expected to be short-lived. Witness the number of minimills following Nucor into some variation of thin-slab casting.

Sustaining success will require ongoing improvements in both technology and products, moving forever upward into more sophisticated, higher value-added markets. And this is the strategy displayed by the companies featured in this book. Achieving this dynamic competitive advantage requires having an organization that anticipates and adapts to change. This is the fundamental underpinning of a sustainable competitive advantage: never being complacent or satisfied with the status quo.

The starting point in building this advantage is a very simple goal: To invest in machines, to invest in people, and to invest in the organization of work so that you get the most out of the resources you command. Industry leaders today may not have set out by realizing and articulating this goal; however, the investments that they made in technology paid off big precisely because they also invested in people and in building a high-performance workplace. In hindsight we recognize the importance of their decisions, but we can also point to them as leading the way to competitiveness for others.

General Application

Few companies embrace the principles that we have described as a whole, consistently, and over extended periods of time. However, those that do so prosper as a result. The failure involved has many sources. Some would lay it to the short-term orientation of management or to corporate functional departments that guard their autonomy jealously. Stifling corporate bureaucracies are at work too, and in some firms this is compounded by the fact that historical success and arrogance has led to resistance or lethargy when new ideas are introduced.[10] Indeed, examples abound of companies that have ceased to be innovative or have ignored their customers. Similarly, one can point to entire U.S. industries that once dominated their markets, only to see their competitiveness overshadowed by more nimble, better-managed foreign companies. These examples extend well beyond steel. One would have to include autos, consumer electronics, machine tools, and textiles, as well as steel, as cases in point.

In every one of these examples the companies involved either forgot or never understood exactly how crucial are the linkages that exist between the leadership of a firm and the firm's technology of production and management. There is no magical solution to achieving the internal conditions necessary for success: It is hard work, and requires a long-term commitment from all levels within the organization. But the elements of competitiveness that we have identified outline an approach that has succeeded for many firms.

U.S. Steel has restructured itself significantly since the early 1980s, doing many of the same things that companies in industries as diverse as financial services, electrical power, automobiles, computers, and retail trade were doing to become more competitive. Inefficient plants were closed, and the focus on markets and customers was sharpened. Layers of organization were removed, and decision making was pushed downward. Accountability was underscored at all levels of the organization. Product quality was increased, and costs were reduced. Importantly, the firm also increased its commitment to its employees through a union agreement that guaranteed a forty-hour workweek and an investment in employee training and employee involvement.

The point here is not what USS has done (others in many industries have taken similar actions), but that USS views these changes as part of an evolving process. In the words of the general manager of one of its plants,

> The future offers more of the same; corporate headquarters will shrink in size; decision making will be driven further and further down into the organization; the hourly workforce will have greater input and assume greater responsibility; cost reduction will be ongoing; growth will occur through increased production and not huge capital expenditures; markets will be served by products with more value-added content; and our focus will continue to be on providing steel that enables our customers to better serve their customers.[11]

The cultural change that started with Tom Graham's leadership in the early 1980s has continued unabated, although the emphasis has shifted as needed to take new circumstances into account.

Many industries are being buffeted by comparable forces of change, although the factors precipitating the change may vary from case to case. The electric utility industry is a case in point. The regulatory framework that governs this industry is being dismantled, and companies are being forced to think and act in a more market-oriented way. This will bring about wrenching change within companies. Their profitability will no longer depend entirely upon rates negotiated with regulatory authorities, but upon their ability to meet the needs of their customers in a more competitive marketplace with costs and service at least as good as their competitors'.[12]

The organizational and cultural changes required in such situations are immense, but in many respects they characterize the requirements in *any* industry that is forced to move from a protected environment to one that is fully open to competition. A cost-plus mentality has to go, and the firm has to accept the fact that its success will depend upon the ability of the company to compete in a world where customers have choice, and cost, quality, and service matter a lot.

Farther afield, take the example of Sears, once a dominant force in the retail trade sector. Along came Kmart and Wal-Mart, and the factors for success in the industry shifted as these discounters captured significant market share from Sears. Sears had to learn to change the way it did business and to change its organization. Later Wal-Mart began to expand into food as it opened mammoth supercenters throughout the country. Retailers and wholesalers in the grocery industry are now having to rethink the way they have traditionally done business. Suddenly information systems technology is becoming more important, human resource management has received greater emphasis, and companies are beginning to rethink their approach to serving markets and customers, and their approach to working with their suppliers.[13]

Many of the principles underscored in this book come into play in responding to challenges like these. Leadership is required to identify the firm's core market and focus investment in technologies that will serve that market well. This vision has to be defined clearly for the firm and communicated throughout the organization. Bottlenecks and imbalances in the operation have to be eliminated in order to raise utilization rates. Complementary investments have to be identified and initiated in the skill and knowledge required of employees. The organization of work has to change; with today's technology this means ensuring that the workforce is flexible and decision making is decentralized. None of this will work well unless the culture of the firm supports the necessary transformation. Employees must share common goals, and they must be highly motivated. The success of a firm in the long haul will turn on whether or not its leadership can carry that off too.

These generalizations provide a helpful road map for how to become or remain competitive in the future. They do not, however, provide the specifics for what GM, or any other company, should do to be a success, because crafting the details requires a thorough understanding of where the company is today in terms of its culture, core competencies, competitive strength, and the like. As we stated in an earlier chapter, there is no "silver bullet," no single best course of action. To pull it all together, dynamic leadership is necessary. Strategies such as total quality management or business process reengineering could be a part of what a company elects to do to improve itself. But these approaches, even if they are totally integrated into the way in which the company conducts its business, are only a part of the answer. More is required.

The critical elements of competitiveness for any firm lie in its leadership, its technology, and its management practices. But the key to achieving and sustaining competitiveness for any firm is to be found in the interrelationships among these elements. In the final analysis, the leadership of a firm must understand how the technology that is required to succeed in particular markets can support and be supported by an organization that brings out the very best in its employees.

Notes

Chapter 1

1. United States Bureau of the Census. *Statistical Abstract of the United States: 1994*, 114th ed. (Washington, D.C.: 1994), Table 656, p. 422.

2. Ibid.

3. American Iron and Steel Institute. *Annual Statistical Report, 1993* (Washington, D.C.: American Iron and Steel Institute, 1994), Table 1B, p. 4; and American Iron and Steel Institute. *Annual Statistical Report, 1984* (Washington, D.C.: American Iron and Steel Institute, 1985), Table 1B, p. 8.

4. Ibid.

5. Tsukasa Furukawa. "Japanese Steel Restructures under the Weight of the Yen." *American Metal Market, Steel Monthly: International Steel* (October 4, 1994), p. 12A, p. 13A; Michael Williams. "Japan Labor System: A Two-Edged Sword." *Wall Street Journal* (November 8, 1994), p. A19; Michiyo Nakamoto. "A Radical Solution for Japan's Double Whammy." *Financial Times* (December 5, 1994), p. 8; and Eamonn Fingleton. "Jobs for Life: Why Japan Won't Give Them Up." *Fortune* (March 20, 1995), pp. 120–25.

6. *The Politics of Steel: Western Europe and the Steel Industry in the Crisis Years, 1974–1984*. Eds. Yves Meny and Vincent Wright (New York: Walter de Gruyter, 1987), pp. 1–110.

7. United States International Trade Commission. *Steel Industry Annual Report, On Competitive Conditions in the Steel Industry and Industry Efforts to Adjust and Modernize*, Publication 2436 (Washington, D.C.: United States International Trade Commission, September 1991), p. 4–14.

8. Jonathan P. Hicks. "An industrial comback story: U.S. is competing again in steel." *New York Times* (March 31, 1992), p. A1.

9. Stephen Baker. "Why Steel Is Looking Sexy." *Business Week* (April 4, 1994), p. 106.

10. Donald F. Barnett and Robert W. Crandall. *Up from the Ashes: The Rise of the Steel Minimill in the United States* (Washington, D.C.: Brookings Institution, 1986), offered one of the first comprehensive discussions of this phenomenon.

11. For example, see American Iron and Steel Institute. *Steel at the Crossroads: The American Steel Industry in the 1980s* (Washington, D.C.: American Iron and Steel Institute, 1980); "Time Runs Out for Steel." *Business Week* (June 13, 1983), pp. 84–94; Donald F. Barnett and Louis Schorsch. *Steel: Upheaval in a Basic Industry* (Cambridge, Mass.: Ballinger Publishing Company, 1983), pp. 3–103; Paul R. Lawrence and Davis Dyer. *Renewing American Industry* (New York: Free Press, 1983), pp. 55–85; Steel Advisory Committee. *Report of the Steel Advisory Committee* (Washington, D.C.: U.S. Department of Commerce and U.S. Department of Labor, December 3, 1984), pp. 45–140; John P. Hoerr. *And the Wolf Finally Came: The Decline of the American Steel Industry*

(Pittsburgh, Penn.: University of Pittsburgh Press, 1988), pp. 1–23; Thomas R. Howell, et al. *Steel and the State: Government Intervention and Steel's Structural Crisis* (Boulder, Colo.: Westview Press, 1988), Chap. 7; Paul A. Tiffany. *The Decline of American Steel: How Management, Labor and Government Went Wrong* (New York: Oxford University Press, 1988); Michael L. Dertouzos, Richard K. Lester, and Robert M. Solow. *Made in America: Regaining the Productive Edge* (Cambridge, Mass.: MIT Press, 1989), pp. 278–87; and James Moss and Anthony Taccone. "Steel in the Developed World: The Evolution of an Industry Continues." Presentation to the Steel Manufacturers' Association, 1994 Annual Member Conference (Washington, D.C., April 20, 1994).

12. Dertouzos, Lester, and Solow. *Made in America*, pp. 278–87.

13. Barnett and Crandall. *Up from the Ashes*, pp. 18–70.

14. United States International Trade Commission. *Steel Industry Annual Report, On Competitive Conditions*, Figure 4–4, p. 4–6, and Table 4–3, pp. 4–7; Donald F. Barnett. "ReInventing the American Steel Industry." Presentation at CAMSI: The Future of Computer Automated Manufacturing in the Steel Industry (May 24, 1993); Richard Fruehan et al. "Manufacturing and Technology Assessment of International Steel Plants." Working Paper, Sloan Steel Project (Pittsburgh, Penn.: Department of Engineering and Materials, Carnegie Mellon University October 1993); Anthony J. DeArdo. "Technology and Competitiveness: Analysis of Bar and Rod Steel Producers." Working Paper, Sloan Steel Project (Pittsburgh, Penn.: Department of Material Science and Engineering, University of Pittsburgh, March 1994); Lawrence Chimerine et al. *Can the Phoenix Survive? The Fall and Rise of the American Steel Industry* (Washington, D.C.: Economic Strategy Institute, June 1994), pp. 85–94; John Holusha. "Why American Steel Is Big Again," *New York Times* (July 21, 1994), p. C1; Marvin B. Lieberman and Douglas R. Johnson. "Comparative Productivity of Japanese and U.S. Steel Producers, 1958–1993." Working Paper, Sloan Steel Project (Los Angeles, Calif.: Anderson Graduate School of Management, University of California at Los Angeles, May 1995); Stephen Baker. "A Real Steelman for USX." *Business Week* (May 15, 1995), p. 47; and John Holusha. "Steel Mini-Mills Could Bring Boom or Blood Bath." *New York Times* (May 30, 1995), pp. C1, C6.

Chapter 2

1. See American Iron and Steel Industry. *Steel at the Crossroads.*

2. Howell et al. *Steel and the State,* pp. 1–14.

3. The complexity and intensity of these political relations are described well in Meny and Wright. *The Politics of Steel.*

4. The historical discussion of the U.S. steel industry draws heavily upon the following sources: William T. Hogan. *Economic History of the Iron and Steel Industry in the United States* (Lexington, Mass.: Lexington Books, 1971); William T. Hogan. *The 1970s: Critical Years for Steel* (Lexington, Mass.: Lexington Books, 1972); Robert W. Crandall. *The U.S. Steel Industry in Recurrent Crisis* (Washington, D.C.: Brookings Institution, 1981); Barnett and Schorsch. *Steel*; William T. Hogan. *World Steel in the 1980s: A Case of Survival* (Lexington, Mass.: Lexington Books, 1983); Lawrence and Dyer. *Renewing American Industry*, Chap. 3; Barnett and Crandall. *Up from the Ashes*; Hoerr. *And the Wolf Finally Came*; Howell et al. *Steel and the State*, Chaps. 2, 7, 8; Mark Reutter. *Sparrows Point: Making Steel—The Rise and Ruin of American Industrial Might* (New York: Summit Books, 1988); Tiffany. *The Decline of American Steel*; William T. Hogan. *Global Steel in the 1990s: Growth or Decline* (Lexington, Mass.: Lexington Books, 1991).

5. See, American Iron and Steel Institute, *Financing Capital Expenditure* (Wash-

ington, D.C.: American Iron and Steel Institute, 1976) for an example of the assumptions underlying industry forecasts during this period.

6. United States International Trade Commission. *Steel Semiannual Monitoring Report, Special Focus: U.S. Industry Conditions*. Publication 2878 (Washington, D.C.: United States International Trade Commission, April 1995), Table 6, p. 20.

7. Barnett and Crandall. *Up from the Ashes*, Table 3–3, p. 44.

8. Marvin Berkowitz and Krishna Mohan. "The Role of Global Procurement in the Value Chain of Japanese Steel." *Columbia Journal of World Business* 12 (1987): 97–110; Krishna Mohan and Marvin Berkowitz. "Raw Materials Procurement Strategy: The Differential Advantage in the Success of Japanese Steel." *Journal of Purchasing and Materials Management* (spring, 1988): 15–22.

9. William Serrin. *Homestead: The Glory and Tragedy of an American Steel Town* (New York: Times Books, 1992).

10. For a discussion of the aftermath of this strike, see Tiffany. *The Decline of American Steel*, pp. 167–84.

11. Hoerr. *And the Wolf Finally Came*, pp. 109–33.

12. American Iron and Steel Institute. *Annual Statistical Report*; Barnett and Crandall. *Up from the Ashes*, Table 3–3, p. 44.

13. Hoerr. *And the Wolf Finally Came*, pp. 297–304.

14. Barnett and Schorsch. *Steel*, pp. 107–39.

15. Howell et al. *Steel and the State*, p. 522.

16. Good discussions of voluntary restraint agreements are contained in American Iron and Steel Institute. *Steel at the Crossroads*; Comptroller General of the United States. *Report to the Congress: New Strategy Required for Aiding Distressed Steel Industry* (Washington, D.C.: U.S. Government Printing Office, 1981); Crandall. *U.S. Steel Industry in Recurrent Crisis*; Robert S. Walters. "U.S. Negotiation of Voluntary Restraint Agreements in Steel, 1984: Domestic Sources of International Economic Diplomacy." *Case Study in International Negotiations* (Pittsburgh, Penn.: Graduate School of Public and International Affairs, University of Pittsburgh, 1987); Committee on Ways and Means, U.S. House of Representatives, 101st Congress, 1st Session. *Background Materials Relating to the Steel Voluntary Restraint Agreement (VRA) Program* (Washington, D.C.: U.S. Government Printing Office, 1989); and Roger S. Ahlbrandt. "Involvement of the U.S. Government in the Steel Industry." Working Paper, Sloan Steel Project (Pittsburgh, Penn.: Katz Graduate School of Business, University of Pittsburgh, August 1992).

17. American Iron and Steel Institute. *Annual Statistical Report, 1992* (Washington, D.C.: American Iron and Steel Institute, 1993), Table 1A, p. 4.

18. U.S. Bureau of the Census. *Statistical Abstract of the United States: 1987*, 107th ed. (Washington, D.C.: 1986), Table 661, p. 389.

19. American Iron and Steel Institute. *Annual Statistical Report, 1992*, Table 2, p. 8; and American Iron and Steel Institute. *Annual Statistical Report, 1984* (Washington, D.C.: American Iron and Steel Institute, 1985), Table 2A, p. 13.

20. Ibid.

21. Howell et al. *Steel and the State*, pp. 510–34; and United States International Trade Commission. *Steel Industry Annual Report 1991*, pp. 1–1, 1–2, 3–22 to 3–25.

22. American Iron and Steel Institute. *Annual Statistical Report, 1992*, Table 1A, p. 4.

23. For statements representing the steel industry's position on this policy, see *Issues Crucial to the Terms of Steel VRA Extension* (Washington, D.C.: American Iron and

Steel Institute, 1989); and American Iron and Steel Institute. *Why Caterpillar's Wrong on VRAs*, which were both issued by the American Iron and Steel Institute in 1989.

24. Caterpillar, Inc. "Steel Import Quotas: Update" (Peoria, Ill.: Caterpillar, Inc., April 1989); and Coalition of American Steel Using Manufacturers. "The Impact of Steel Quotas on American Competitiveness" (Washington, D.C.: Coalition of American Steel Using Manufacturers, January 20, 1989).

25. The White House, Office of the Press Secretary. "Statement by the President" (July 25, 1989); and Clyde H. Farnsworth. "Steel Import Quotas Extended While U.S. Seeks Subsidy Curbs." *New York Times* (July 26, 1989), p. A1.

26. Rose Gutfeld, and Dana Milbank. "U.S. Steel Firms Get Early Boost in Import Fight." *New York Times* (August 11, 1992), p. A2.

27. Keith Bradsher. "U.S. Imposes Heavy Tariffs on Steel from 19 Countries." *New York Times* (January 18, 1993), p. A1.

28. Keith Bradsher. "U.S. Rules Out Many Tariffs on Imported Steel." *New York Times* (July 28, 1993), p. C1; United States International Trade Commission. *Volume I: Determinations and Views of the Commission*, Publication 2664 (Washington, D.C.: United States International Trade Commission, August 1993); and United States International Trade Commission. *Volume II: Information Obtained in the Investigation*, Publication 2664 (Washington, D.C.: United States International Trade Commission, August, 1993).

29. For background reading on the European steel crisis and the role played by government, see Meny and Wright. *The Politics of Steel*; Susan H. Houseman. *Industrial Restructuring with Job Security: The Case of European Steel* (Cambridge, Mass.: Harvard University Press, 1991); and Roger S. Ahlbrandt and Frank Giarratani. "The European Communities: Responding to the Crisis in the Global Steel Industry." Working Paper, Sloan Steel Project (Pittsburgh, Penn.: Katz Graduate School of Business, University of Pittsburgh, August 1992).

30. Computations made by the authors; see Ahlbrandt and Giarratani. "The European Communities," p. 15. Data obtained from the *Eurostat Iron and Steel Yearbook*, 1977, 1980, 1985, and 1991 (Luxemburg: Statistical Office of the European Communities). The figures were complied for Germany, France, Italy, Netherlands, Belgium, Luxemburg, United Kingdom, Ireland, and Denmark.

31. See World Steel Dynamics. *Financial Dynamics of 61 International Steelmakers, Core Report Y* (January 1988), pp. 2–101; and *Core Report LL* (September 1990), pp. 2–99, reproduced in United States International Trade Commission. *Steel Industry Annual Report*, 1991, pp. 2–28.

32. Ahlbrandt and Giarratani. "The European Communities," p. 17.

33. Bulletin of the European Communities 9, 11(1976): 14–16; *Bulletin of the European Communities*, 13, 10 (1980): 7–12; Loukas Tsoukalis and Robert Strauss. "Community Policies on Steel, 1974–1982: A Case of Collective Management," in Meny and Wright. *Politics of Steel*; and Howell et al. *Steel and the State*, Chap. 3.

34. Thomas Grunert. "Decision making Processes in the Steel Crisis Policy of the EEC: Neocorporatist or Integrationist Tendencies?" in Meny and Wright. *Politics of Steel*.

35. *Bulletin of the European Communities* 13, 10 (1980): 7–12; Tsoukalis and Strauss. "Community Policies" in Meny and Wright *Politics of Steel*; and Howell et al. *Steel and the State*.

36. Ahlbrandt and Giarratani. "The European Communities," p. 15.

37. *Bulletin of the European Communities* 14, 6 (1981): 17–19.

38. *Bulletin of the European Communities* 15, 9 (1982): 19.

39. *Bulletin of the European Communities* 15, 11 (1982): 18.

40. See the articles in Meny and Wright. *Politics of Steel*, describing state policy for steel in the U.K., France, Italy, Germany, and Belgium; and Houseman. *Industrial Restructuring*.

41. EC Commission. *Report from the Commission to the Council on the Application of the Rules on Aid to the Steel Industry, 1984–85*, Com (86), 235 final (August 6, 1986); reproduced in Howell et al. *Steel and the State*, p. 83; and Ahlbrandt and Giarratani. "The European Communities," p. 18.

42. *Bulletin of the European Communities* 18, 9 (1985): 26.

43. EC Commission. *Rules on Aid to the Steel Industry*. See also Ahlbrandt and Giarratani. "The European Communities," p. 21.

44. The development of the Japanese steel industry and the role played by MITI is well covered in the following: Kiyoshi Kawahito. *Japanese Steel Industry with an Analysis of the U.S. Steel Import Problem* (New York: Praeger, 1972); Chalmers Johnson. *MITI and the Japanese Miracle: The Growth of Industrial Policy, 1925–1975* (Stanford, Calif.: Stanford University Press, 1982); Leonard Lynn. *How Japan Innovates: A Comparison with the U.S. in the Case of Oxygen Steelmaking* (Boulder, Colo.: Westview Press 1982); Howell et al. *Steel and the State*, Chap. 4; and *Changing Patterns of International Rivalry: Some Lessons from the Steel Industry*, eds. Etsuo Abe and Yoshitaka Suzuki (Tokyo: University of Tokyo Press, 1991).

45. In Chapter 3, we will discuss the minimill cost advantage in more detail.

46. United States Bureau of the Census. *Statistical Abstract of the United States: 1987*, 107th ed. (Washington, D.C.: 1987), Table 661, p. 389; and United States Bureau of the Census. *Statistical Abstract of the United States: 1994*, 114th ed. (Washington, D.C.: 1994), Table 656, p. 422.

47. American Iron and Steel Institute. *Annual Statistical Report, 1975* (Washington, D.C.: American Iron and Steel Institute, 1976), Table 6, p. 21; and American Iron and Steel Institute. *Annual Statistical Report, 1993*, Table 6, p. 15.

48. American Iron and Steel Institute. *Annual Statistical Report, 1984*, Table 1A, p. 8; and American Iron and Steel Institute. *Annual Statistical Report, 1993*, Table 1A, p. 4.

49. American Iron and Steel Institute and The Steel Manufacturers' Association. *Steel: A National Resource for the Future* (Washington, D.C.: American Iron and Steel Institute and The Steel Manufacturers Association, May 1995), p. 6; and Marvin B. Lieberman and Douglas R. Johnson. "Comparative Productivity of Japanese & US Steel Producers, 1958–1993." Working Paper, Sloan Steel Project (Los Angeles, Calif.: Anderson Graduate School of Management, University of California at Los Angeles, May 1995).

50. We refer here to the reported or estimated capacity of furnaces in U.S. steel plants. For technical and economic reasons, furnace capacity may exceed actual production capability in the industry by 15 percent or more.

51. For this description, we use a rather broad definition of "specialty" steel plants, one that includes a wide range of alloy steel production. Often this category is limited to stainless steel and alloy tool steel only.

52. American Iron and Steel Institute. *Annual Statistical Report, 1993*, Table 7, p. 16.

53. United States International Trade Commission. *Steel Semiannual Monitoring, Special Focus: U.S. Industry Conditions* Publication 2878 (Washington, D.C.: United States International Trade Commission, April 1995), Table 13, p. 32.

54. United States International Trade Commission. *Steel Semiannual Monitoring*

Report, Special Focus: Steel Product Quality and Customer Service, Publication 2807 (Washington, D.C.: United States International Trade Commission, September 1994), pp. 17–27.

55. United States International Trade Commission. *Steel Semiannual Monitoring Report, Steel Product Quality*, p. 18.

Chapter 3

1. Our estimates of the market share accounted for by end users reflect the prevailing opinions of industry experts. See, for example, The WEFA Group. "Major Steel End-Markets," *Steel Market Outlook* (Bala Cynwyd, Penn.: The WEFA Group, 1991), pp. 1.9–1.22. In fact, as reported in the American Iron and Steel Institute. *Annual Statistical Report, 1993* (Washington, D.C.: American Iron and Steel Institute, 1994), p. 21, more than one-fourth of all steel shipments go to Steel Service Centers where they are further processed and shipped to end users such as the construction and automotive industries.

2. American Iron and Steel Institute. *Annual Statistical Report 1992*, (Washington, D.C.: American Iron and Steel Institute, 1993), Table 1B, p. 4.

3. Ibid., Table 1A, p. 4.

4. United States International Trade Commission. *Steel: Semiannual Monitoring Report, Special Focus: U.S. Industry Conditions*, Publication 2655 (Washington, D.C.: United States International Trade Commission, June 1993), Table E–5, p. E–6.

5. Andrew Hill, and Andrew Baxter "This Could Be the Last Time: EC Support for Stricken Steelmakers Will Depend on Industry Restructuring." *Financial Times* (November 24, 1992), p. 20; Andrew Hill. "EC Maps a Path for the Steel Industry's Contraction," *Financial Times* (February 15, 1993), p. 2; "Down the Drain: Plans to Restructure Europe's Steel Industry Are in Trouble," *Economist* (December 11, 1993), p. 73; Andrew Baxter. "EU Steel Industry Moves to Next Phase of Battle." *Financial Times* (December 20, 1993), p. 2; Andrew Baxter. "British Steel Doubts Mettle of Brussels Deal," *Financial Times* (January 12, 1994), p. 15; and Emma Tucker and Andrew Baxter. "EU Steel Rescue Plan Is Dead," *Financial Times* (May 21–22, 1994), p. 2.

6. In Chapter 9, we shall describe these ventures more fully and explain their importance in the context of the competitive struggle for the flat product market.

Chapter 4

1. The restructuring of United States Steel was documented through interviews with numerous executives, including David M. Roderick (CEO of USX Corp., 1987); Thomas C. Graham (CEO of USS, 1983–90), in a series of interviews and discussions during the period 1990–95; Thomas Usher (President of USS, 1992); John Goodwin (General Manager of USS Gary Works, 1992); Thomas Sterling (Vice President, Human Relations, USS Group, 1995); James Kutka (Director, Human Resources, USS Group, 1995) Paul Wilhelm (President of USS, 1995); and Reuben Perin (Executive Vice President of USS, 1995). In-depth site visits were made to the Gary Works and the Mon Valley Works, internal documents were made available showing the improvements at the plant level for the Gary Works from 1982–92, and publicly available information was analyzed, including annual reports and 10Ks (annual reports required to be submitted to the Securities and Exchange Commission).

2. The growth of Nucor was documented through a review of publicly available data, site visits to its Crawfordsville and Hickman plants and interviews with execu-

tives, including John Correnti (President, 1993, 1995), Keith Busse (Vice President and General Manager of Crawfordsville plant, 1992), and Rodney Mott (Vice President and General Manager of Hickman plant, 1995).

3. Review of 10Ks, various years.

4. See Chapter 2, Table 2.1 for data on steelmaking capacity by industry subsector in 1974 and 1994.

5. The growth of the minimill sector is discussed in the following: Barnett and Crandall. *Up from the Ashes*; Mark Russell. "Small Steelmakers Find Profitable Niches." *Wall Street Journal* (January 8, 1987), p. 6; George J. McManus. "On Your Mark, Get Set, Grow." *Iron Age* (May 1991), pp. 14–19; Terence Paré. "The Big Threat to Big Steel's Future." *Fortune* (July 15, 1991), pp. 104–8; Locker Associates. "The Minimill Sector: Future Directions." Presentation for the USWA Executive Board Meeting, April 3, 1992; "America's Steel Industry: Protection's Stepchild." *Economist* (May 16, 1992), pp. 97–98; Martin Dickson. "A Little Lesson for Big Steel." *Financial Times* (August 7, 1992), p. 12; Dana Milbank. "Big Steel Is Threatened by Low Cost Rivals, Even in Japan, Korea." *Wall Street Journal* (February 2, 1993), p. 1; and *American Metal Market, Steel Monthly: Electric Furnace Steel* (February 18, 1993).

6. Patricia Beeson and Frank Giarratani. "Spatial Aspects of Capacity Change by Integrated Steel Producers in the United States." Working Paper, Sloan Steel Project (Pittsburgh, Penn.: University of Pittsburgh, 1995).

7. Discussions of Nucor are contained in the following: Michael Schroeder and Walecia Konrad. "Nucor: Rolling Right into Steel's Big Time." *Business Week* (November 19, 1990), pp. 76–79; Anne Lobel Armel. "Competition Thins Out as Iverson Casts Nucor's Lot." *Iron Age* (August 1991), pp. 20–22; Kenneth F. Iverson. "Effective Leadership: The Key Is Simplicity." In *The Quest for Competitiveness,* eds. Y. K. Shelty and V. M. Buehler. (New York: Quorum Books, 1991), pp. 285–94; Richard Preston. *American Steel* (New York: Prentice Hall, 1991); and Martin Dickson. "Ideas from Indiana: Recent Face of Steel." *Financial Times* (July 28, 1992), p. 10.

8. Interview with John Correnti, President of Nucor, October 8, 1993.

9. Preston. *American Steel*, p. 74.

10. Nucor's thin-slab casting decision and the subsequent start-up at the Crawfordsville plant are fully described in Preston. *American Steel.*

11. Correnti interview, 1993; interview with Rodney Mott, Vice President and General Manager of Nucor's Hickman plant, June 27, 1995.

12. Correnti interview, 1993.

13. As of November 1995, the Nucor iron carbide plant was over one year behind schedule and operating at only 20 percent of its capacity. Major engineering modifications have been required to correct the problems encountered. These setbacks do not speak generally to the commercial viability of iron carbide or other forms of DRI; rather, they are particular to the process adopted in the Nucor venture. Other firms have invested in alternative processes that are now coming on line or are in the proposal stage. There is no doubt that scrap metal substitutes from virgin iron will be important in the years to come.

14. Len Boselovic. "Steelmakers Rush to Darwinian Conclusion." *Pittsburgh Post-Gazette* (November 27, 1994), p. B1.

15. Michael Marley. "Mini-Mills Cook up New Melt Mix Recipes." *American Metal Market, Electric Furnace Steel* (February 24, 1994), pp. 4A, 5A.

16. For example, see "The Worldwide Steel Industry: Reshaping to Survive." *Business Week* (August 20, 1984), pp. 150–54; Howell et al. *Steel and the State,* Chaps. 1,

2, and 7; Dertouzos, Lester and Solow. *Made in America,* pp. 278–87; and Stephen Baker et al. "Fat City for Big Steel: But How Long Will the Glory Days Last?" *Business Week* (August 22, 1994), pp. 24–25.

17. Hoerr. *And the Wolf Finally Came*, Chaps. 13, 14, and 16.

18. Interview with David M. Roderick, CEO of USX Corp., 1987.

19. Computed from the company's annual reports and 10Ks, 1982, 1983; also see William C. Symonds et al. "The Toughest Job in Business: How They're Remaking U.S. Steel." *Business Week* (February 25, 1985), pp. 50–56.

20. Michael Schroeder. "This 'Barracuda' Is Still on the Attack." *Business Week* (January 20, 1992), pp. 96–97; Dana Milbank. "Graham, Known for Steel Turnarounds, Is Selected as Chief of Amco Venture." *Wall Street Journal* (June 2, 1992), p. B9; Erle Norton,. "Chairman AK Steel Tries to Shake Off Tag of 'Operating Man'." *Wall Street Journal* (November 25, 1994), p. 1; John Holusha. "Having Done It All in Steel, He's on Top at Last." *New York Times* (February 12, 1995), p. F8.

21. Interview with Thomas C. Graham, July 24, 1992.

22. Graham interview, 1992.

23. Graham interview, 1992.

24. Graham interview, 1992.

25. Graham interview, 1992.

26. Graham interview, 1992.

27. Graham interview, 1992.

28. Graham interview, 1992.; review of internal data provided for the Gary Works.

29. Graham interview, 1992; review of annual reports and 10Ks, various years.

30. Interview with Thomas C. Graham, June 2, 1994.

31. Graham interview, 1994.

32. Graham interview, 1994.

33. According to the American Iron and Steel Institute, *Annual Statistical Report, 1993* (Washington, D.C.: American Iron and Steel Institute, 1994), Table 10, p. 25, domestic shipments in this market were 44 million tons in 1993.

Chapter 5

1. Interview with Thomas C. Graham, June 2, 1994.

2. Interview with John Correnti, October 8, 1993.

3. Graham interview, 1994.

4. Graham interview, 1994.

5. Graham interview, 1994.

6. Graham interview, 1994.

7. Graham interview, 1994.

8. Graham interview, 1994; internal data provided by the company.

9. Correnti interview, 1993.

10. Correnti interview, 1993; interview with Keith Busse, Vice President and General Manager of Nucor's Crawfordsville plant, February 26, 1992; and interview with Rodney Mott, Vice President and General Manger of Nucor's Hickman Plant, June 27, 1995.

11. Correnti interview, 1993.

12. Information on Oregon Steel Mills was obtained from the company's annual reports, publicly available information, and interviews conducted with executives, managers, supervisors, and production workers. See Roger S. Ahlbrandt et al. "Case

Study of Oregon Steel Mills, Inc." Working Paper, Sloan Steel Project (Portland, Oreg.: School of Business Administration, Portland State University, November 1993); and Frank Haflich. "Pipe Deal Boosts Oregon." *American Metal Market* 101, 226 (November 22, 1993), p. 4.

13. Interview with Thomas B. Boklund, Chairman and CEO, and Vicki Tagliafico, Manager of Strategic Planning and Investor Relations, Oregon Steel Mills, Inc., August 3, 1992.

14. Boklund and Tagliafico interview, 1992.

15. Boklund and Tagliafico interview, 1992.

16. Boklund and Tagliafico interview, 1992.

17. Boklund and Tagliafico interview, 1992.

18. Data provided by the company, August 14, 1995.

19. Correnti interview, 1993.

20. Mott interview, 1995.

21. Correnti interview, 1993.

22. Correnti interview, 1993.

23. Information on Birmingham Steel was obtained from the company's annual reports, publicly available information, and interviews with executives, managers, supervisors, and production workers. See Roger S. Ahlbrandt. "Case Study of Birmingham Steel Corporation." Working Paper, Sloan Steel Project (Pittsburgh, Penn.: Katz Graduate School of Business, University of Pittsburgh, September 1992).

24. Interview with James A. Todd, Jr., Chairman and CEO, Birmingham Steel Corporation, August 5, 1992.

25. Ahlbrandt. "Case Study of Birmingham Steel Corporation."

26. Ahlbrandt. Information provided by the company, August 16, 1995.

27. Busse interview, 1992; Preston. *American Steel*, pp. 140–50.

28. Busse interview, 1992; Preston. *American Steel*, pp. 18–20.

29. Correnti interview, 1993.

30. Preston. *American Steel*, pp. 103–266.

31. Interview with Kevin Ratliff, in Ahlbrandt et al. "Case Study of Oregon Steel Mills," p. 31. Emphasis added.

32. Todd interview, 1992.

33. Review of the company's annual reports.

34. Interview with Joe Corvin, President, Oregon Steel, July 26, 1993; also Bryan Berry. "A Construction Engineer Helps Build a Steelmaker." *New Steel* (August 1994), pp. 14–22.

35. Ahlbrandt. "Case Study of Birmingham Steel Corporation"; and *Birmingham Steel Corporation Prospectus and American Steel and Wire Corporation Proxy Statement* (October 25, 1993), issued with respect to the merger of the two companies.

36. Boklund and Tagliafico interview, 1992.

Chapter 6

1. U.S. International Trade Commission. *Steel Semiannual Monitoring Report, Special Focus: U.S. Industry Conditions*, Publication 2878 (Washington, D.C.: U.S. International Trade Commission, April 1995), Table 12, p. 31.

2. R. J. Fruehan et al. "Manufacturing and Technology Assessment of International Steel Plants." Working Paper, Sloan Steel Project (Pittsburgh, Penn.: Department of Engineering and Materials, Carnegie-Mellon University, 1993); a revised version of the manuscript is published in *Iron and Steelmaker* (January 1994): pp. 25–31.

3. "Japanese Steel Programs with U.S. Steelmakers: Technical Cooperation/Joint Ventures Aid U.S. Restructuring; Provide Quality Steel for American Manufacturers' Efficiency." (New York: Japan Steel Information Center, April 1991).

4. James P. Womack, Daniel T. Jones, and Daniel Roos. *The Machine that Changed the World* (New York: Rawson Associates, 1990).

5. Paul Milgrom, and John Roberts. "The Economics of Modern Manufacturing: Technology, Strategy, and Organization." *American Economics Review* 80, 3 (1990): 511–28.

6. Some related themes are presented in Edith T. Penrose. *The Theory of the Growth of the Firm* (White Plains, N.Y.: M. E. Sharpe, 1980). See especially, pp. 43–87.

7. Interview with Thomas B. Boklund, Chairman and CEO, and Vicki Tagliafico, Manager of Strategic Planning and Investor Relations, Oregon Steel Mills, Inc., August 3, 1992.

8. Interviews with John F. Kaloski, General Manager, Mon Valley Works, U.S. Steel, and John McCluskey, Manager, Employee Relations, Mon Valley Works, U.S. Steel, August 25, 1995.

9. Adam Smith. *The Wealth of Nations: Representative Selections.* Ed. Bruce Mazlish. (New York: Bobbs-Merrill Company, 1961).

10. Alan S. Blinder. "Introduction." In *Paying for Productivity: A Look at the Evidence*, ed. Alan S. Blinder (Washignton, D.C.: Brookings Institution, 1990), p. 13.

11. Casey Ichniowski, Katherine Shaw, and Giovanna Prennushi. "The Effect of Human Resource Management Practices on Productivity." Working Paper, Sloan Steel Project (Pittsburgh, Penn.: Graduate School of Industrial Administration, Carnegie Mellon University, 1994).

12. McClusky interview, 1995; a broader discussion of this issue is contained in Carrie R. Leana, and Robert Albright. "First-Line Supervisors' Receptivity to Employee Involvement: An Examination of Competing Predictors." Unpublished manuscript (Pittsburgh, Penn.: Katz Graduate School of Business, University of Pittsburgh, 1994).

13. Roger Ahlbrandt et al. "Case Study of Oregon Steel Mills, Inc." Working Paper, Sloan Steel Project (Portland, Oreg.: School of Business Administration, Portland State University, November 1993), p. 6.

14. Interview with Rodney Mott, Vice President and General Manger of Nucor's Hickman Plant, June 27, 1995.

15. Interview with James A. Todd, Jr., Chairman and CEO, Birmingham Steel Corporation, August 5, 1992.

16. Interview with John Correnti, President of Nucor, October 8, 1993.

17. Interview with Keith Busse, Vice President and General Manager of Crawfordsville plant, February 26, 1992; Correnti interview, 1993; and Mott interview, 1995.

18. Correnti interview, 1993.

Chapter 7

1. For background on the European experience, see Meny and Wright, *The Politics of Steel;* Howell et al. *Steel and the State,* Chap. 3; and Houseman. *Industrial Restructuring with Job Security.*

2. This chapter does not focus on the leadership role played by top managers. From the discussion, however, it will be obvious that the extensive changes that occurred within these companies would not have been possible unless a strong committed leadership had been at work.

3. This section is based upon review of annual reports and other published infor-

mation on British Steel, plc; a site visit to British Steel's corporate offices and its Port Talbot Works; interviews with P. J. K. Ferguson, Director of Personnel, E. W. Denham, Director of Corporate Planning, and R. D. Thompson, Manager of Competitive Analysis (July 4, 1994); interviews with John M. Waine, Manager, Personnel, Integrated Works at Port Talbot, and a number of other managers at the plant (July 5, 1994).

4. Meny and Wright. *The Politics of Steel.* pp. 13–14.
5. Ferguson interview, 1994.
6. Data from various annual reports.
7. Data reported in annual reports; also Ferguson interview, 1994.
8. Dana Milbank. "British Steel's Great Pain Turns to Profit." *Wall Street Journal* (December 28, 1993), p. A4; Andrew Baxter. "British Steel Doubts Mettle of Brussels Deal." *Financial Times* (January 12, 1994), p. 15; Andrew Baxter. "British Steel Sees Future in the East." *Financial Times* (June 21, 1994), p. 26; Andrew Baxter. "British Steel Core Is More Than Before." *Financial Times* (February 15, 1995), p. 15.
9. Ferguson interview, 1994.
10. Waine interview, 1994, and review of internal plant documents.
11. Ferguson interview, 1994.
12. "Memorandum of Agreement made on 15th May, 1980, between the British Steel Corporation, Port Talbot Works, and the Joint Trade Unions, Port Talbot Works, with regard to Major Changes in Principles, Working Practices, Job Reductions, Improved Productivity and Reduced Costs at Port Talbott Works," p. 3, an internal document made available by plant management, July 5, 1994.
13. Waine interview, 1994.
14. Denham interview, 1994.
15. Denham interview, 1994; Ferguson interview, 1994; Waine interview, 1994.
16. Review of annual reports; Denham interview, 1994; Ferguson interview, 1994; and Ted Denham. "Expanding the British Steel Empire." *American Metal Market, Steel Monthly* (September 21, 1995), p. 14.
17. Denham interview, 1994; Ferguson interview, 1994.
18. Ferguson interview, 1994; Waine interview, 1994.
19. Ferguson interview, 1994; Waine interview, 1994.
20. Ferguson interview, 1994; Waine interview, 1994.
21. Ferguson interview, 1994; Waine interview, 1994.
22. Ferguson interview, 1994; Waine interview, 1994.
23. Ferguson interview, 1994; Waine interview, 1994.
24. This section is based upon an analysis of publicly available information, a site visit to Co-Steel's Sheerness plant; Hugh E. Billot. "Fast Moving Cultural Change Programmes at Co-Steel Sheerness." Paper presented at the meeting of the American Iron and Steel Society, Williamsburg, Va., November 1993; and interviews with John W. Clayton, Works Director, and Hugh E. Billot, Personnel Director, July 7, 1994.
25. Billot interview, 1994; Clayton interview, 1994.
26. Billot interview, 1994; Clayton interview, 1994.
27. Billot interview, 1994; Clayton interview, 1994.
28. Billot. "Fast Moving Cultural Change Programmes."
29. Billot interview, 1994; Clayton interview, 1994.
30. Billot interview, 1994; Clayton interview, 1994.
31. Billot interview, 1994; Clayton interview, 1994.
32. Billot interview, 1994; Clayton interview, 1994.
33. Billot interview, 1994; Clayton interview, 1994.

34. Billot interview, 1994; Clayton interview, 1994.
35. Billot interview, 1994; Clayton interview, 1994.
36. Billot interview, 1994; Clayton interview, 1994.
37. Billot interview, 1994; Clayton interview, 1994.
38. Billot interview, 1994; Clayton interview, 1994.
39. Billot interview, 1994; Clayton interview, 1994.
40. Billot interview, 1994; Clayton interview, 1994.
41. Billot interview, 1994; Clayton interview, 1994.
42. Billot interview, 1994; Clayton interview, 1994.
43. Billot interview, 1994; Clayton interview, 1994.
44. Billot interview, 1994; Clayton interview, 1994.

Chapter 8

1. The information for Nippon Steel was obtained from annual reports and other publicly available documents; interviews with Tomokatsu Kobayashi, Vice President, Nippon Steel U.S.A., Inc. (1992, 1994); and interviews with managers in the Corporate Planning Division and Labor Division at Nippon's Tokyo headquarters (1995).

2. The information for Tokyo Steel was obtained from annual reports and other publicly available documents; a site visit to the Company's Okayama plant, and interviews with the managers (1995); and an interview with Masanari Iketani, President of the company (1995).

3. Historical background for this section can be found in Kawahito. *Japanese Steel Industry*; Johnson. *MITI and the Japanese Miracle*; Lynn. *How Japan Innovates*; Howell et al. *Steel and the State*, Chap. 4; and *Changing Patterns of International Rivalry*. Eds. Abe and Suzuki, pp. 166–244.

4. International Iron and Steel Institute. *World Steel in Figures, 1992* (Brussels, Belgium: International Iron and Steel Institute, 1992), p. 18.

5. Ibid.

6. Merton J. Peck, Richard C. Levin, and Akira Goto. "Picking Losers: Public Policy toward Declining Industries in Japan." *Journal of Japanese Studies* 13, 1 (1987): 79–123; interview with Tomakatsu Kobayashi, Vice President and Secretary, Nippon Steel U.S.A., Inc., January 15, 1992; interview with Chikashi Morimoto, Assistant Manager, Tokyo Personnel Section, Sumitomo Metal Industries Ltd., January 30, 1995; interview with Shigefumi Kawamoto, Manager, Labor Planning Department, Labor Division, Nippon Steel, February 2, 1995.

7. Kobayashi interview, 1992.

8. Japan Iron and Steel Federation, various publications; statistics reproduced in Nippon Steel Corporation. *Basic Facts about Nippon Steel, 1993* (Tokyo: Nippon Steel, 1993), p. 118.

9. Nippon Steel. *Annual Reports* (various years).

10. Michiyo Nakamoto. "Japan's Steelmakers under Pressure." *Financial Times* (May 31, 1994), p. 20; "Japanese Steel: Blasted Furnaces." *Economist* (June 4, 1994), pp. 63–64; Tsukasa Furukawa. "Japanese Steel Restructures under the Weight of the Yen." *American Metal Market, Steel Monthly: International Steel* (October 4, 1994), pp. 12A, 13A; Tsukasa Furukawa. "Japan Steel Takes Pounding but Remains Hopeful." *American Metal Market* (June 7, 1995), p. 5.

11. Interview with Tomokatsu Kobayashi, Vice President and Secretary, Nippon Steel U.S.A., Inc., October 28, 1994; and interviews with Takahiro Hino, General Manager, Corporate Planning Division, Toshihiko Ono, General Manager, Corporate Planning Division, and Shigefumi Kawamoto, Manager, Labor Planning Division,

Nippon Steel Corporation, January 30, 1995.

12. See annual reports of Nippon Steel Corp., NKK Corp., Kawasaki Steel Corp., Sumitomo Metal Industries, Ltd., and Kobe Steel Corp., 1993, 1994, 1995; also Furukawa. "Japan Steel Takes Pounding."

13. Based on statistics in Japan Iron and Steel Federation. *The Steel Industry of Japan, 1994* (Tokyo: Japan Iron and Steel Federation, 1994).

14. Michiyo Nakamoto. "Mitsubishi Breaks Ranks to Buy Steel from Korea." *Financial Times* (December 9, 1994), p. 1.

15. Yoshiro Sasaki, Executive Vice President of Nippon Steel Corp., in Furukawa. "Japanese Steel Restructures," p. 12A; also see Michiyo Nakamoto. "A Radical Solution for Japan's Double Whammy." *Financial Times* (December 5, 1994), p. 8.

16. Furukawa. "Japanese Steel Restructures."

17. Based on Kobayashi interview, 1994; Hino interview, 1995; Kawamoto interview, 1995; Ono interview, 1995.

18. Hino interview, 1995; Kawamoto interview 1995; Ono interview, 1995.

19. Hino interview, 1995; Kawamoto interview, 1995; Ono interview, 1995.

20. Hino interview, 1995; Kawamoto interview, 1995; Ono interview, 1995.

21. Morimoto interview, 1995.

22. Hino interview, 1995; Kawamoto interview, 1995; Ono interview, 1995.

23. Hino interview, 1995; Kawamoto interview, 1995; Morimoto interview, 1995; Ono interview, 1995.

24. Hino interview, 1995; Kawamoto interview, 1995; Ono interview, 1995.

25. Kobayashi interview, 1994; Hino interview, 1995; Ono interview, 1995; Kawamoto interview, 1995.

26. Kobayashi interview, 1994; Hino interview, 1995; Ono interview, 1995.

27. Kobayashi interview, 1994.

28. Furukawa. "Japanese Steel Restructures"; and Tsukasa Furukawa. "In Japan It's Godzilla vs. Mini-Godzilla." *American Metal Market, Steel Monthly: Mini-Mill Steel* (December 6, 1994), pp. 10A-11A.

29. Furukawa. "Japanese Steel Restructures," p. 13A.

30. Furukawa. "Godzilla vs. Mini-Godzilla," p. 10A.

31. Hino interview, 1995; Kawamoto interview, 1995; Ono interview, 1995.

32. Interview with Masanari Iketani, President, Tokyo Steel Manufacturing Company, February 1, 1995.

33. Iketani interview, 1995.

34. Furukawa. "Japanese Steel Restructures."

35. Hino interview, 1995; Kawamoto interview, 1995; Ono interview, 1995.

36. Iketani interview, 1995.

37. Hino interview, 1995; Kawamoto interview, 1995; Ono interview, 1995.

38. Iketani interview, 1995.

39. Iketani interview, 1995.

40. Iketani interview, 1995.

41. Iketani interview, 1995.

42. Iketani interview, 1995.

43. For a good discussion of the *Kaizen* process, see Masaaki Imai. *Kaizen: The Key to Japan's Competitive Success* (New York: McGraw Hill, 1986); and Hedrick Smith. *Rethinking America* (New York: Random House, 1995), pp. 23–26.

44. Hino interview, 1995; Kawamoto interview, 1995; Morimoto interview, 1995; Ono interview, 1995.

45. Hino interview, 1995; Kawamoto interview, 1995; Morimoto interview,

1995; Ono interview, 1995; and interview with Hitoshi Toita, General Manager, General Affairs Department, Okayama Plant, Tokyo Steel Manufacturing Co., Ltd., January 31, 1995.

46. Toita interview, 1995.

47. Toita interview, 1995.

48. Nippon Steel Corporation, *Basic Facts about Nippon Steel, 1994* (Tokyo: Nippon Steel, 1994), p. 21.

49. Hino interview, 1995; Kawamoto interview, 1995; Morimoto interview.

Chapter 9

1. No new integrated steel plant has been built in the United States in the past twenty-five years. Lacking recent experience, estimates of the capital costs associated with new plant construction are necessarily crude. However, an estimate of $3 to 5 billion to construct an integrated plant capable of producing 3 million ton per year would be reasonable. New EAF thin-slab plants in the range of 1 to 2 million tons per year are being built for a total cost of $400 to $500 million. See, for example, R. J. Fruehan, H. W. Paxton, F. Giarratani, and L. Lave. *Future Steelmaking Industry and Its Technologies*. Idaho National Engineering Laboratory, Report INEL 95/0046 (December 1994).

2. See John Schriefer. "Coping with the Coke Crunch." *New Steel* (July 1995), pp. 26–31. Schriefer reports that coke production has decreased from a rate of 65 million tons per year in 1973 to a rate of only 22 million tons per year today.

3. A state-of-the-art blast furnace uses 500 kg of coke per metric tonne (1,000 kg) of hot metal. Coal injection can replace half the coke requirement.

4. Corex plants have been planned or constructed in Korea, South Africa, and India. The process involved produces very substantial excess energy in the form of off-gases. If Corex is to enjoy widespread application, this excess energy will have to harnessed for other uses such as power generation or the production of direct reduced iron.

5. The processes that are near commercialization are AISI Direct Steelmaking (United States), the DIOS process (Japan), and HIsmelt (Austria and Japan). The exact form of the technology that will emerge in the marketplace is uncertain, and it could well be some combination of those currently being developed.

6. Nippon Steel-Mitsubishi, Usinor Sacilor-Thyssen, and BHP-IHI have run fairly large pilot plants casting 25 ton heats up to about 1.5M in width.

7. Currently, the scrap used in the steel industry and in foundries breaks down as about 30 percent home scrap, 20 percent prompt scrap, and 50 percent obsolete scrap. We can expect home and prompt scrap to continue to decrease in availability, and the use of obsolete scrap to continue to increase as a result.

8. Georgetown Steel has also proposed to build a DRI plant capable of producing over one million tons per year in the Gulf region of the United States with partners who would provide capital and take a share of the DRI.

9. Jo Isenberg-O'Loughlin. "Gallatin Enters the Thin-Slab Sweepstakes." *33 Metal Producing* (September 1995), p. 34.

10. Ted Denham. "Expanding the British Steel Empire." *American Metal Market, Steel Monthly* (September 21, 1995), p. 14.

11. United States International Trade Commission. *Steel Semiannual Monitoring Report, Special Focus: U.S. Industry Conditions*. Publication 2878 (Washington, D.C.: United States International Trade Commission, April 1995), Table 6, p. 20.

12. Ibid.

13. Ibid.

14. Ibid.

15. Nippon Steel and Usinor Sacilor have strip casters for stainless steel that could be commercialized in the next two or three years.

16. R. J. Fruehan. "Critical Research and the Impact of Emerging Technologies in Steelmaking." *Metallurgical Processes for the Twenty-First Century* (Warrendale, Penn.: TMS, 1994).

17. See United States International Trade Commission. *Steel Industry Annual Report: On Competitive Conditions in the Steel Industry and Industry Efforts to Adjust and Modernize.* Publication 2316 (Washington, D.C.: United States International Trade Commission, September 1990), for a description of the major research partnerships ongoing during the 1980s. Cooperative projects are also described in United States International Trade Commission. *Steel: Semiannual Monitoring Report. Special Focus: U.S. Industry Conditions.* Publication 2655 (Washington, D.C.: United States International Trade Commission, June 1993), pp. 25–28.

18. United States International Trade Commission. *Steel Semiannual Monitoring Report. Special Focus: U.S. Industry Conditions.* Publication 2759 (Washington, D.C.: United States International Trade Commission, April 1994), p. 10.

19. American Iron and Steel Institute. *Annual Statistical Report, 1993.* (Washington, D.C.: American Iron and Steel Institute, 1994), Table 10, p. 25 and Table 18, p. 50.

20. For example, USS is investing in vacuum degassing at its Edgar Thomson plant in Pittsburgh in order to make it possible to produce ultralow carbon steels.

Chapter 10

1. Tom Peters. *Liberation Management: Necessary Disorganization for the Nanosecond Nineties* (New York: Alfred A. Knopf, 1992).

2. For reading related to learning organizations see C. Argyris and D. Schön. *Organizational Learning* (Reading, Mass.: Addison-Wesley, 1978); R. H. Hays, S. C. Wheelwright, and K. B. Clark. *Dynamic Manufacturing: Creating the Learning Organization* (New York: Free Press, 1988); R. H. Kilmann, T. J. Covin and Associates. *Corporate Transformation: Revitalizing Organizations for a Competitive World* (San Francisco: Jossey-Bass, 1988); B. Levitt, and J. G. March. "Organizational Learning." *Annual Review of Sociology*, 14 (1988):319–40; P. M. Senge. *The Fifth Discipline: The Art and Practice of the Learning Organization.* (New York: Doubleday/Currency, 1990); R. H. Kilmann, I. Kilmann, and Associates. *Making Organizations Competitive: Enhancing Networks and Relationships across Traditional Boundaries.* (San Francisco: Jossey-Bass, 1991); David A. Garvin. "Building a Learning Organization." *Harvard Business Review*, (July/August 1993): 78–91.

3. Interview with Rodney Mott, Vice President and General Manager of Nucor's Hickman plant, June 27, 1995.

4. "Australian Rolling Mill Set for NW." *Oregonian* (June 6, 1995), p. E1.

5. Interviews with P. J. K. Ferguson, Director of Personnel; and E. W. Denham, Director of Corporate Planning, British Steel, Port Talbot Works, July 4, 1994.

6. Interviews with Takahiro Hino, General Manager, Corporate Planning Division, and Toshihiko Ono, General Manager, Corporate Planning Division, Nippon Steel Corporation, January 30, 1995.

7. Gloria T. LaRue. "Tuscaloosa Gets Go-Ahead." *American Metal Market* (October 14, 1994), p. 1.

8. Ibid.

9. Kerri J. Selland. "3 Steelmakers Link Forces in Mini-Mill." *American Metal Market* (December 20, 1994), p. 1; and Stephen Baker. "Cheap Steel Doesn't Come Quick." *Business Week* (July 17, 1995), p. 35.

10. Suzanne Berger et al. "Toward a New Industrial America. "*Scientific American* (June 1989), pp. 39–47.

11. Interview with John F. Kaloski, General Manager, Mon Valley Works, U.S. Steel, August 25, 1995.

12. Agis Salpukas. "How a Staid Electric Company Became a Renegade." *New York Times* (December 12, 1993), p F10; and Fred Buckman (President and CEO of PacifiCorp). "Competitiveness: Challenges Facing the Electric Utility Industry." Presentation, School of Business Administration, Portland State University, January 26, 1995.

13. John Huey. "Sam Walton in His Own Words."*Fortune* (June 29, 1992), pp. 98–106; Carol J. Loomis. "Dinosaurs?" *Fortune* (May 3, 1993), pp. 36–42; Mark A. Husson. "Wal-Mart, Company Report."(New York: JP Morgan, September 7, 1994); and Susan Chandler. "Where Sears Wants America to Shop Now." *Business Week* (June 12, 1995), p. 39.

Selected Bibliography

Abe, Etsuo, and Yoshitaka Suzuki, eds. *Changing Patterns of International Rivalry: Some Lessons from the Steel Industry*. Tokyo: University of Tokyo Press, 1991.

American Iron and Steel Institute. *Steel at the Crossroads: The American Steel Industry in the 1980s*. Washington, D.C.: American Iron and Steel Institute, 1980.

American Iron and Steel Institute, and The Steel Manufacturers' Association. *Steel: A National Resource for the Future.* Washington, D.C.: American Iron and Steel Institute and The Steel Manufacturers' Association, May 1995.

Barnett, Donald F., and Robert W. Crandall. *Up from the Ashes: The Rise of the Steel Minimill in the United States.* Washington, D.C.: Brookings Institution, 1986.

Barnett, Donald F., and Louis Schorsch. *Steel: Upheaval in a Basic Industry.* Cambridge, Mass.: Ballinger Publishing Company, 1983.

Berkowitz, Martin, and Krishna Mohan. "The Role of Global Procurement in the Value Chain of Japanese Steel." *Columbia Journal of World Business* 12 (1987): 97–110.

Blinder, Alan S. "Introduction." In *Paying for Productivity: A Look at the Evidence*, ed. Alan S. Blinder. Washington, D.C.: Brookings Institution, 1990.

Chimerine, Lawrence et al. *Can the Phoenix Survive? The Fall and Rise of the American Steel Industry.* (Washington, D.C.: Economic Strategy Institute, June 1994.

Committee on Ways and Means, U.S. House of Representatives, 101st Congress, 1st Session. *Background Materials Relating to the Steel Voluntary Restraint Agreement (VRA) Program*. Washington, D.C.: U.S. Government Printing Office, 1989.

Comptroller General of the United States. *Report to the Congress: New Strategy Required for Aiding Distressed Steel Industry*. Washington, D.C.: U.S. Government Printing Office, 1981.

Crandall, Robert W. *The U.S. Steel Industry in Recurrent Crisis*. Washington, D.C.: Brookings Institution, 1981.

Dertouzos, Michael L., Richard K. Lester, and Robert M. Solow. *Made in America: Regaining the Productive Edge.* Cambridge, Mass.: MIT Press, 1989.

Fruehan, Richard. "Critical Research and the Impact of Emerging Technologies in Steelmaking." *Metallurgical Processes for the Twenty-First Century*. Warrendale, Penn.: TMS, 1994.

Hoerr, John P. *And the Wolf Finally Came: The Decline of the American Steel Industry.* Pittsburgh, Penn.: University of Pittsburgh Press, 1988.

Hogan, William T. *Economic History of the Iron and Steel Industry in the United States.* Lexington, Mass.: Lexington Books, 1971.

———. *Global Steel in the 1990s: Growth or Decline.* Lexington, Mass.: Lexington Books, 1991.

———. *The 1970s: Critical Years for Steel.* Lexington, Mass.: Lexington Books, 1972.

———. *World Steel in the 1980s: A Case of Survival.* Lexington, Mass.: Lexington Books, 1983.

Houseman, Susan H. *Industrial Restructuring with Job Security: The Case of European Steel.* Cambridge, Mass.: Harvard University Press, 1991.

Howell, Thomas R. et al. *Steel and the State: Government Intervention and Steel's Structural Crisis.* Boulder, Colo.: Westview Press, 1988.

Imai, Masaaki. *Kaizen: The Key to Japan's Competitive Success.* New York: McGraw Hill, 1986.

Iverson, Kenneth F. "Effective Leadership: The Key Is Simplicity." In *The Quest for Competitiveness,* eds. Y. K. Shelty and V. M. Buehler. New York: Quorum Books, 1991.

Johnson, Chalmers. *MITI and the Japanese Miracle: The Growth of Industrial Policy, 1925–1975.* Stanford, Calif.: Stanford University Press, 1982.

Kawahito, Kiyoshi. *Japanese Steel Industry with an Analysis of the U.S. Steel Import Problem.* New York: Praeger, 1972.

Lawrence, Paul R., and Davis Dyer. *Renewing American Industry.* New York: Free Press, 1983.

Levitt, B., and J. G March. "Organizational Learning." *Annual Review of Sociology* 14 (1988): 319–40.

Lynn, Leonard. *How Japan Innovates: A Comparison with the U.S. in the Case of Oxygen Steelmaking.* Boulder, Colo.: Westview Press, 1982.

Meny, Yves, and Vincent Wright, eds. *The Politics of Steel: Western Europe and the Steel Industry in the Crisis Years, 1974–1984.* New York: Walter de Gruyter, 1987.

Milgrom, Paul, and John Roberts. "The Economics of Modern Manufacturing: Technology, Strategy, and Organization." *American Economics Review* 80, 3 (1990): 511–28.

Mohan, Krishna, and Marvin Berkowitz. "Raw Materials Procurement Strategy: The Differential Advantage in the Success of Japanese Steel." *Journal of Purchasing and Materials Management* (spring 1988): 15–22.

Peck, Merton J., Richard C. Levin, and Akira Goto. "Picking Losers: Public Policy Toward Declining Industries in Japan." *Journal of Japanese Studies* 13, 1 (1987): 79–123.

Penrose, Edith T. *The Theory of the Growth of the Firm.* White Plains, N.Y.: M. E. Sharpe 1980.

Preston, Richard. *American Steel.* New York: Prentice Hall, 1991.

Reutter, Mark. *Sparrows Point: Making Steel—The Rise and Ruin of American Industrial Might.* New York: Summit Books, 1988.

Serrin, William. *Homestead: The Glory and Tragedy of an American Steel Town.* New York: Times Books, 1992.

Smith, Adam. *The Wealth of Nations, Representative Selections*, ed. Bruce Mazlish. New York: Bobbs-Merrill Company, 1961.

Smith, Hedrick. *Rethinking America.* New York: Random House, 1995.

Steel Advisory Committee. *Report of the Steel Advisory Committee.* Washington, D.C.: U.S. Department of Commerce and U.S. Department of Labor, December 3, 1984.

Tiffany, Paul A. *The Decline of American Steel: How Management, Labor, and Government Went Wrong.* New York: Oxford University Press, 1988.

United States International Trade Commission. *Steel Industry Annual Report: Financial Dynamics of 61 International Steelmakers, Core Report Y,* January 1988. Washington, D.C.: United States International Trade Commission, 1991.

United States International Trade Commission. *Steel Industry Annual Report: On Competitive Conditions in the Steel Industry and Industry Efforts to Adjust and Modernize.*

Publication 2316. Washington, D.C.: United States International Trade Commission, September 1990.

United States International Trade Commission. *Steel Industry Annual Report: On Competitive Conditions in the Steel Industry and Industry Efforts to Adjust and Modernize.* Publication 2436. Washington, D.C.: United States International Trade Commission, September, 1991.

Womack, James P., Daniel T. Jones, and Daniel Roos. *The Machine that Changed the World.* New York: Rawson Associates, 1990.

Sloan Steel Project Papers

Ahlbrandt, Roger S. "Case Study of Birmingham Steel Corporation." Working Paper, Sloan Steel Project. Pittsburgh, Penn.: Katz Graduate School of Business, University of Pittsburgh, September 1992.

———. "Involvement of the U.S. Government in the Steel Industry." Working Paper, Sloan Steel Project. Pittsburgh, Penn.: Katz Graduate School of Business, University of Pittsburgh, August 1992.

Ahlbrandt, Roger S. et al. "Case Study of Oregon Steel Mills, Inc." Working Paper, Sloan Steel Project. Portland, Oreg.: School of Business Administration, Portland State University, November 1993.

Ahlbrandt, Roger S., and Frank Giarratani. "The European Communities: Responding to the Crisis in the Global Steel Industry." Working Paper, Sloan Steel Project. Pittsburgh, Penn.: Katz Graduate School of Business, University of Pittsburgh, August 1992.

Beeson, Patricia, and Frank Giarratani. "Spatial Aspects of Capacity Change by Integrated Steel Producers in the United States." Working Paper, Sloan Steel Project. Pittsburgh, Penn.: University of Pittsburgh, 1995.

DeArdo, Anthony J. "Technology and Competitiveness: Analysis of Bar and Rod Steel Producers." Working Paper, Sloan Steel Project. Pittsburgh, Penn.: Department of Material Science and Engineering, University of Pittsburgh, March 1994.

Fruehan, Richard et al. "Manufacturing and Technology Assessment of International Steel Plants." Working Paper, Sloan Steel Project. Pittsburgh, Penn.: Department of Engineering and Materials, Carnegie Mellon University, October 1993.

Ichniowski, Casey, Katherine Shaw, and Giovanna Prennushi. "The Effect of Human Resource Management Practices on Productivity." Working Paper, Sloan Steel Project. Pittsburgh, Penn.: Graduate School of Industrial Administration, Carnegie Mellon University, 1994.

Lieberman, Marvin B. and Douglas R. Johnson, "Comparative Productivity of Japanese and U.S. Steel Producers, 1958–1993." Working Paper, Sloan Steel Project. Los Angeles, Calif.: Anderson Graduate School of Management, University of California at Los Angeles, May 1995.

Index